ENGINEERING HYDROLOGY

Edited by
June Blyth Orillana Royol

Delve Publishing

Engineering Hydrology

Edited by June Blyth Orillana Royol

Edition 2018

Delve Publishing

Published by

Delve Publishing LLC

708 3rd Avenue, 6th Floor

New York

NY 10801

United States of America

www.delvepublishing.com

ISBN: 978-1-68095-686-3 (Hardback)

Printed in United States of America

Library of Congress Control Number: 2016962000

About the Editor

June Blyth Orillana Royol

June Blyth Royol completed her study at Negros Oriental State University. An accredited Contractors' and Consultants' Materials Engineer I since April 2015. Published a book entitled "Optimum Water-Cement Ratio for 50% Recycled Concrete as Coarse Aggregate for Building Construction". She is currently working as a Licensed Civil Engineer at the Department of Public Works and Highways.

Table of Contents

5 INFILTRATION OF HYDROLOGY

6 WATER QUALITY

7 SURFACE RUNOFF AND ROUTING

8 GROUNDWATER

9 WATERSHED MANAGEMENT

10 DRAINAGE BASIN

11 HYDROLOGICAL TRANSPORT MODEL

List of Abbreviations

AEM	Analytic Element Method
AGU	American Geophysical Union
BEM	Boundary Element Method
BOD	Biochemical Oxygen Demand
COD	Chemical Oxygen Demand
CSO	Combined Sewer Overflow
CUAHSI	Universities for the Advancement of Hydrologic Science, Inc.
DRH	Direct Runoff Hydrograph
EPA	Environmental Protection Agency
ESC	Erosion and Sediment Controls
ET	Evapotranspiration
FDA	Food and Drug Administration
FEM	Finite Element Method
FZDZ	Freezing Drizzle
FZRA	Freezing Rain
GIS	Geographic Information Systems
GSA	Geological Society of America
IAHS	International Association of Hydrological Sciences
IHP	International Hydrological Programme
IWFM	Integrated Water Flow Model
IWMI	International Water Management Institute
LID	Low Impact Development
MCS	Mesoscale Convective System

ODS	Ozone Depleting Substances
ORP	Oxygen Reduction Potential
PDE	Partial Differential Equation
PET	Potential Evapotranspiration
SG	Snow Grains
SSO	Sanitary Sewer Overflow
SWM	Stanford Watershed Model
SWMM	Storm Water Management Model
TDS	Total Dissolved Solids
THMC	Thermo-Hydro-Mechanical-Chemical
UH	Unit Hydrograph

Preface

Hydrology deals with the origin, occurrence, circulation, distribution, the physical and chemical properties of water and its interaction with living organisms. Hydrology is an essential field of science since everything from tiny organisms to individuals to societies to the whole of civilization - depends so much on water. Hydrological engineering, also called water resources engineering, is a civil engineering specialty offered at both the undergraduate and graduate levels. Hydrological engineering is chiefly concerned with the flow and storage of water. Topics commonly covered include urban drainage, water supply, wastewater treatment, river management and coastal protection. Hydrological engineering also focuses on preventing floods and lessening the effects of floods, droughts and other natural disasters. Water is an essential resource that is required by all life on Earth. Studying the movement, availability, and quality of water are the jobs of a hydrologist. More specifically hydrologists study the chemical properties, biological interactions, and the physical processes that govern the water cycle. The water cycle or hydrologic cycle is a process by which water is continuously cycled around the earth. This happens through different pathways and at different rates but the central concepts remain the same. Water evaporates from the ocean, condenses as clouds, moves over land, and precipitates. From there it can enter ground water, evaporate again, or enter a stream or lake. It will eventually find its way back to the ocean either by falling as precipitation, flowing with a river, or by moving ever so slowly with ground water. The hydrologic cycle is also a process that transfers heat energy. Heat is transported pole ward by water being evaporated and then condensing which releases heat. Without the water cycle the climate would be much more frigid and areas away from the equator would be much less habitable. Hydrology is more significant these days because we plan ahead of time to deal with extremes (scarcity of water leading to droughts, and overflowing of water bodies leading to floods). This text introduces students to the key concepts and methods in physical and engineering hydrology.

Content Coverage

Chapters One and *Two* provide an overview of hydrology and hydrogeology, respectively. Hydrogeology deals with the flow of water through aquifers and other shallow porous media.

Chapters Three and *Four* provide coverage of evapotranspiration and precipitation, separately. Evapotranspiration is a significant water loss from drainage basins, whereas precipitation is a major component of the water cycle, and is responsible for depositing the fresh water on the planet.

Chapters Five and *Six* are infiltration engineering and water quality, correspondingly. Infiltration is the process by which water on the ground surface enters the soil, whereas human factors and ergonomics is concerned with the "fit" between the user, equipment and their environments, whereas water quality is a measure of the condition of water relative to the requirements of one or more biotic species and or to any human need or purpose.

Chapters Seven and *Eight* present about surface runoff strategies and groundwater. Surface runoff is a crucial part of the water system. Typically, groundwater is thought of as water flowing through shallow aquifers, but, in the technical sense, it can also contain soil moisture, permafrost (frozen soil), immobile water in very low permeability bedrock, and deep geothermal or oil formation water.

Chapters Nine and *Ten* cover watershed management and drainage basin, respectively. Watershed management provides a framework for integrated decision-making to help: assess the nature and status of the watershed; identify watershed issues; define and re-evaluate short and long-term objectives, actions and goals; assess benefits and costs; and implement and evaluate action. The drainage basin acts as a funnel by collecting all the water within the area covered by the basin and channeling it to a single point.

Chapter Eleven focuses an approach on the hydrological transport model used to simulate river or stream flow and calculate water quality parameters.

CHAPTER 1

Hydrology

Chapter Contents

- **Introduction**
- **Branches**
- **Applications**
- **History**
- **Themes**

Hydrology is the study of water and its properties, scientific laws, and distribution. Hydrologists study surface water and groundwater and their interactions, which are part of the water cycle.

Surface water is freshwater we see above ground. It includes streams, lakes, ponds and wetlands. Groundwater is freshwater that lies beneath the earth's surface. Most groundwater is stored in aquifers. Aquifers are areas underground where rock and soil are very porous.

Many features of streams and rivers undergo constant change. Throughout a single day we may observe changes in water level, salinity, water flow speed, discharge (the amount of water passing a point on the river in a given amount of time), and sediment load (the amount of soil and gravel suspended in the water). These changes are caused by water input into the river. Rain and melting snow increase the amount of freshwater in rivers. Rain falls directly into rivers, and rainwater and melted snow on land surrounding the river runs downhill into the river.

INTRODUCTION

Hydrology is the scientific study of the movement, distribution, and quality of water on Earth and other planets, including the water cycle, water resources and environmental watershed sustainability. A practitioner of hydrology is a hydrologist, working within the fields of earth or environmental science, physical geography, geology or civil and environmental engineering.

Hydrology subdivides into surface water hydrology, groundwater hydrology (hydrogeology), and marine hydrology. Domains of hydrology include hydrometeorology, surface hydrology, hydrogeology, drainage-basin management and water quality, where water plays the central role.

Oceanography and meteorology are not included because water is only one of many important aspects within those fields.

Hydrological research can inform environmental engineering, policy and planning.

Water covers 70% of the Earth's surface.

BRANCHES

- Chemical hydrology is the study of the chemical characteristics of water.
- Ecohydrology is the study of interactions between organisms and the hydrologic cycle.
- Hydrogeology is the study of the presence and movement of groundwater.
- Hydroinformatics is the adaptation of information technology to hydrology and water resources applications.
- Hydrometeorology is the study of the transfer of water and energy between land and water body surfaces and the lower atmosphere.
- Isotope hydrology is the study of the isotopic signatures of water.
- Surface hydrology is the study of hydrologic processes that operate at or near Earth's surface.
- Drainage basin management covers water-storage, in the form of reservoirs, and flood-protection.

- Water quality includes the chemistry of water in rivers and lakes, both of pollutants and natural solutes.

APPLICATIONS

- Determining the water balance of a region.
- Determining the agricultural water balance.
- Designing riparian restoration projects.
- Mitigating and predicting flood, landslide and drought risk.
- Real-time flood forecasting and flood warning.
- Designing irrigation schemes and managing agricultural productivity.
- Part of the hazard module in catastrophe modeling.
- Providing drinking water.
- Designing dams for water supply or hydroelectric power generation.
- Designing bridges.
- Designing sewers and urban drainage system.
- Analyzing the impacts of antecedent moisture on sanitary sewer systems.
- Predicting geomorphologic changes, such as erosion or sedimentation.
- Assessing the impacts of natural and anthropogenic environmental change on water resources.
- Assessing contaminant transport risk and establishing environmental policy guidelines.

HISTORY

Hydrology has been a subject of investigation and engineering for millennia. For example, about 4000 BC the Nile was dammed to improve agricultural productivity of previously barren lands. Mesopotamian towns were protected from flooding with high earthen walls. Aqueducts were built by the Greeks and Ancient Romans, while the history of China shows they built irrigation and flood control works. The ancient Sinhalese used hydrology to build complex irrigation works in Sri Lanka, also known for invention of the Valve Pit which allowed construction of large reservoirs, anicuts and canals which still function.

Marcus Vitruvius, in the first century BC, described a philosophical theory of the hydrologic cycle, in which precipitation falling in the mountains infiltrated the Earth's surface and led to streams and springs in the lowlands. With adoption of a more scientific approach, Leonardo da Vinci and Bernard Palissy independently reached an accurate representation of the hydrologic cycle. It was not until the 17th century that hydrologic variables began to be quantified.

Pioneers of the modern science of hydrology include Pierre Perrault, Edme Mariotte and Edmund Halley. By measuring rainfall, runoff, and drainage area, Perrault showed that rainfall was sufficient to account for flow of the Seine. Marriotte combined velocity and river cross-section measurements to obtain discharge, again in the Seine. Halley showed that the evaporation from the Mediterranean Sea was sufficient to account for the outflow of rivers flowing into the sea.

Advances in the 18th century included the Bernoulli piezometer and Bernoulli's equation, by Daniel Bernoulli, and the Pitot tube, by Henri Pitot. The 19th century saw development in groundwater hydrology, including Darcy's law, the Dupuit-Thiem well formula, and Hagen-Poiseuille's capillary flow equation.

Rational analyses began to replace empiricism in the 20th century, while governmental agencies began their own hydrological research programs. Of particular importance were Leroy Sherman's unit hydrograph, the infiltration theory of Robert E. Horton, and C.V. Theis's aquifer test/equation describing well hydraulics.

Since the 1950s, hydrology has been approached with a more theoretical basis than in the past, facilitated by advances in the physical understanding of hydrological processes and by the advent of computers and especially geographic information systems (GIS).

THEMES

The central theme of hydrology is that water circulates throughout the Earth through different pathways and at different rates. The most vivid image of this is in the evaporation of water from the ocean, which forms clouds. These clouds drift over the land and produce rain. The rainwater flows into lakes, rivers, or aquifers. The water in lakes, rivers, and aquifers then either evaporates back to the atmosphere or eventually flows back to the ocean, completing a cycle. Water changes its state of being several times throughout this cycle.

The areas of research within hydrology concern the movement of water between its various states, or within a given state, or simply quantifying the amounts in these states in a given region. Parts of hydrology concern developing methods for directly measuring these flows or amounts of water, while others concern modelling these processes either for scientific knowledge or for making prediction in practical applications.

Groundwater

Ground water is water beneath Earth's surface, often pumped for drinking water. Groundwater hydrology (hydrogeology) considers quantifying groundwater flow and solute transport. Problems in describing the saturated zone include the characterization of aquifers in terms of flow direction, groundwater pressure and, by inference, groundwater depth.

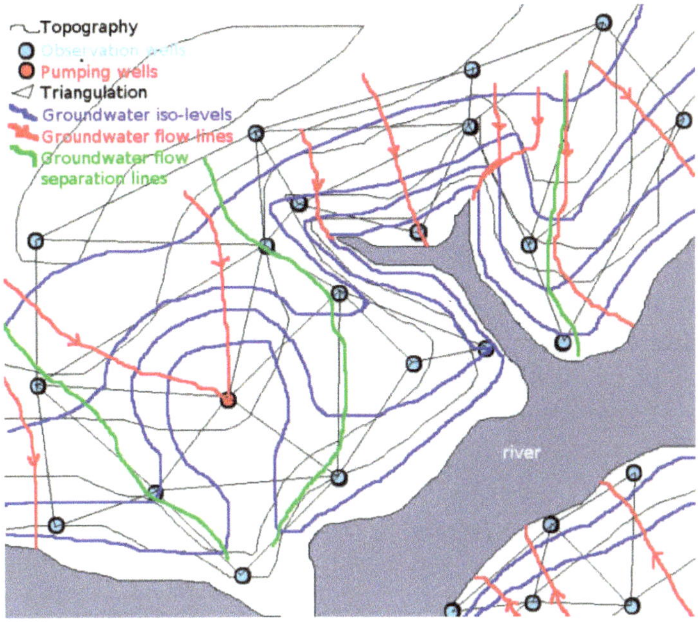

Building a map of groundwater contours

Measurements here can be made using a piezometer. Aquifers are also described in terms of hydraulic conductivity, storativity and transmisivity. There are a number of geophysical methods for characterising aquifers. There are also problems in characterizing the vadose zone (unsaturated zone).

Infiltration

Infiltration is the process by which water enters the soil. Some of the water is absorbed, and the rest percolates down to the water table. The infiltration capacity, the maximum rate at which the soil can absorb water, depends on several factors. The layer that is already saturated provides a resistance that is proportional to its thickness, while that plus the depth of water above the soil provides the driving force (hydraulic head). Dry soil can allow rapid infiltration by capillary action; this force diminishes as the soil becomes wet. Compaction reduces the porosity and the pore sizes. Surface cover increases capacity by retarding runoff, reducing compaction and other processes. Higher temperatures reduce viscosity, increasing infiltration.

Soil Moisture

Soil moisture can be measured in various ways; by capacitance probe, time domain reflectometer or Tensiometer. Other methods include solute sampling and geophysical methods.

Surface Water Flow

Hydrology considers quantifying surface water flow and solute transport, although the treatment of flows in large rivers is sometimes considered as a distinct topic of hydraulics or hydrodynamics. Surface water flow can include flow both in recognizable river channels and otherwise. Methods for measuring flow once water has reached a river include the stream gauge, and tracer techniques. Other topics include chemical transport as part of surface water, sediment transport and erosion.

One of the important areas of hydrology is the interchange between rivers and aquifers. Groundwater/surface water interactions in streams and aquifers can be complex and the direction of net water flux (into surface water or into the aquifer) may vary spatially along a stream channel and over time at any particular location, depending on the relationship between stream stage and groundwater levels.

Precipitation and Evaporation

In some considerations, hydrology is thought of as starting at the land-atmosphere boundary and so it is important to have adequate knowledge of both precipitation and evaporation. Precipitation can be measured in various ways: disdrometer for precipitation characteristics at a fine time scale; radar for cloud properties, rain rate estimation, hail and snow detection; rain gauge for routine accurate measurements of rain and snowfall; satellite for rainy area identification, rain rate estimation, land-cover/land-use, and soil moisture, for example.

Evaporation is an important part of the water cycle. It is partly affected by humidity, which can be measured by a sling psychrometer. It is also affected by the presence of snow, hail and ice and can relate to dew, mist and fog. Hydrology considers evaporation of various forms: from water surfaces; as transpiration from plant surfaces in natural and agronomic ecosystems. A direct measurement of evaporation can be obtained using Simon's evaporation pan.

Detailed studies of evaporation involve boundary layer considerations as well as momentum, heat flux and energy budgets.

Remote Sensing

Remote sensing of hydrologic processes can provide information on locations where *in situ* sensors may be unavailable or sparse. It also enables observations over large spatial extents. Many of the variables constituting the terrestrial water balance, for example surface water storage, soil moisture, precipitation, evapotranspiration, and snow and ice, are measurable using remote sensing at various spatial-temporal resolutions and accuracies. Sources of remote sensing include land-based sensors, airborne sensors and satellite sensors which can capture microwave, thermal and near-infrared data or use lidar, for example.

Water Quality

In hydrology, studies of water quality concern organic and inorganic compounds, and both dissolved and sediment material. In addition, water quality is affected by the interaction of dissolved oxygen with organic material and various chemical transformations that may take place. Measurements of water quality may involve either in-situ methods, in which analyses take place on-site, often automatically, or laboratory-based analyses and may include microbiological analysis.

Integrating Measurement and Modelling

- Budget analyses
- Parameter estimation
- Scaling in time and space
- Data assimilation
- Quality control of data – see for example Double mass analysis

Prediction

Observations of hydrologic processes are used to make predictions of the future behavior of hydrologic systems (water flow, water quality). One of the major current concerns in hydrologic research is "Prediction in Ungauged Basins" (PUB), i.e. in basins where no or only very few data exist.

Statistical Hydrology

By analyzing the statistical properties of hydrologic records, such as rainfall or river flow, hydrologists can estimate future hydrologic phenomena. When making assessments of how often relatively rare events will occur, analyses are made in terms of the return period of such events. Other quantities of interest include the average flow in a river, in a year or by season.

These estimates are important for engineers and economists so that proper risk analysis can be performed to influence investment decisions in future infrastructure and to determine the yield reliability characteristics of water supply systems. Statistical information is utilized to formulate operating rules for large dams forming part of systems which include agricultural, industrial and residential demands.

Modeling

Hydrological models are simplified, conceptual representations of a part of the hydrologic cycle. They are primarily used for hydrological prediction and for understanding hydrological processes, within the general field of scientific modeling. Two major types of hydrological models can be distinguished:

- Models based on data. These models are black box systems, using mathematical and statistical concepts to link a certain input (for instance rainfall) to the model output (for instance runoff). Commonly used techniques are regression, transfer functions, and system identification. The simplest of these models may be linear models, but it is common to deploy non-linear components to represent some general aspects of a catchment's response without going deeply into the real physical processes involved. An example of such an aspect is the well-known behavior that a catchment will respond much more quickly and strongly when it is already wet than when it is dry.
- Models based on process descriptions. These models try to represent the physical processes observed in the real world. Typically, such models contain representations of surface runoff, subsurface flow, evapotranspiration, and channel flow, but they can be far more complicated. These models are known as deterministic hydrology models. Deterministic hydrology models can be subdivided into single-event models and continuous simulation models.

Recent research in hydrological modeling tries to have a more global approach to the understanding of the behavior of hydrologic systems to make better predictions and to face the major challenges in water resources management.

Transport

Water movement is a significant means by which other material, such as soil, gravel, boulders or pollutants, are transported from place to place. Initial input to receiving waters may arise from a point source discharge or a line source or area source, such as surface runoff. Since the 1960s rather complex mathematical models have been developed, facilitated by the availability of high speed computers. The most common pollutant classes analyzed are nutrients, pesticides, total dissolved solids and sediment.

Organizations

Intergovernmental Organizations

- International Hydrological Programme (IHP)

International Research Bodies

- International Water Management Institute (IWMI)
- UNESCO-IHE Institute for Water Education

National Research Bodies

- Centre for Ecology and Hydrology – UK
- Centre for Water Science, Cranfield University, UK
- Seawag – aquatic research, ETH Zürich, Switzerland

- Institute of Hydrology, Albert-Ludwigs-University of Freiburg, Germany
- United States Geological Survey – Water Resources of the United States
- NOAA's National Weather Service – Office of Hydrologic Development, USA
- US Army Corps of Engineers Hydrologic Engineering Center, USA
- Hydrologic Research Center, USA
- NOAA Economics and Social Sciences, USA
- University of Oklahoma Center for Natural Hazards and Disasters Research, USA
- National Hydrology Research Centre, Canada
- National Institute of Hydrology, India

National and International Societies

- Geological Society of America (GSA) – Hydrogeology Division
- American Geophysical Union (AGU) – Hydrology Section
- National Ground Water Association (NGWA)
- American Water Resources Association
- Consortium of Universities for the Advancement of Hydrologic Science, Inc. (CUAHSI)
- International Association of Hydrological Sciences (IAHS)
- Statistics in Hydrology Working Group (subgroup of IAHS)
- German Hydrological Society (DHG: Deutsche Hydrologische Gesellschaft)
- Italian Hydrological Society (SII-IHS) – http://www.sii-ihs.it
- Nordic Association for Hydrology
- British Hydrological Society
- Russian Geographical Society (Moscow Center) – Hydrology Commission
- International Association for Environmental Hydrology
- International Association of Hydrogeologists

Basin- and Catchment-Wide Overviews

- Connected Waters Initiative, University of New South Wales – Investigating and raising awareness of groundwater and water resource issues in Australia
- Murray Darling Basin Initiative, Department of Environment and Heritage, Australia

Research Journals

- Hydrological Processes, ISSN 1099-1085 (electronic) 0885-6087 (paper), John Wiley & Sons

- Hydrology Research, ISSN 0029-1277, IWA Publishing (formerly Nordic Hydrology)
- Journal of Hydroinformatics, ISSN 1464-7141, IWA Publishing
- Journal of Hydrologic Engineering, ISSN 0733-9496, ASCE Publication
- Journal of Hydrology
- Water Research
- Water Resources Research
- Hydrological Sciences Journal - Journal of the International Association of Hydrological Sciences (IAHS) ISSN 0262-6667 (Print), ISSN 2150-3435 (Online)

REFERENCES

1. "What is hydrology and what do hydrologists do?". USA.gov. U.S. Geological Survey. Retrieved 7 October 2015.
2. Vereecken, H.; Kemna, A.; Münch, H. M.; Tillmann, A.; Verweerd, A. (2006). "Aquifer Characterization by Geophysical Methods". Encyclopedia of Hydrological Sciences. John Wiley & Sons. doi:10.1002/0470848944.hsa154b. ISBN 0-471-49103-9.
3. Wilson, L. Gray; Everett, Lorne G.; Cullen, Stephen J. (1994). Handbook of Vadose Zone Characterization & Monitoring. CRC Press. ISBN 978-0-87371-610-9.
4. Reddy, P. Jaya Rami (2007). A textbook of hydrology (Reprint. ed.). New Delhi: Laxmi Publ. ISBN 9788170080992.
5. Tang, Q.; Gao, H.; Lu, H.; Lettenmaier, D. P. (6 October 2009). "Remote sensing: hydrology". Progress in Physical Geography. 33 (4): 490–509. doi:10.1177/0309133309346650.
6. "International Hydrological Programme (IHP)". IHP. Retrieved 8 June 2013.
7. "International Water Management Institute (IWMI)". IWMI. Retrieved 8 March 2013.
8. "UNESCO-IHE Institute for Water Education". UNIESCO-IHE. Retrieved 8 March 2013.
9. "CEH Website". Centre for Ecology & Hydrology. Retrieved 8 March 2013.
10. "Cranfield Water Science Institute". Cranfield University. Retrieved 8 March 2013.
11. "Eawag aquatic research". Swiss Federal Institute of Aquatic Science and Technology. 25 January 2012. Retrieved 8 March 2013.
12. "Professur für Hydrologie". University of Freiburg. 23 February 2010. Retrieved 8 March 2013.
13. "Water Resources of the United States". USGS. 4 October 2011. Retrieved 8 March 2013.
14. "Office of Hydrologic Development". National Weather Service. NOAA. 28 October 2011. Retrieved 8 March 2013.

15. "Hydrologic Engineering Center". US Army Corps of Engineers. Retrieved 8 March 2013.
16. "Hydrologic Research Center". Hydrologic Research Center. Retrieved 8 March 2013.
17. "NOAA Economics and Social Sciences". NOAA Office of Program Planning and Integration. Retrieved 8 March 2013.
18. "Center for Natural Hazard and Disasters Research". University of Oklahoma. 17 June 2008. Retrieved 8 March 2013.
19. "National Hydrology Research Centre (Saskatoon, SK)". Environmental Science Centres. Environment Canada. Retrieved 8 March 2013.
20. "National Institute of Hydrology (Roorkee), India". NIH Roorkee. Retrieved 1 August 2015.
21. "Hydrogeology Division". The Geological Society of America. 10 September 2011. Retrieved 8 March 2013.
22. "Welcome to AGU's Hydrology (H) Section". American Geophysical Union. Retrieved 8 March 2013.
23. "National Ground Water Association". Retrieved 8 March 2013.
24. "American Water Resources Association". 2 January 2012. Retrieved 8 March 2013.
25. "CUAHSI". Retrieved 8 March 2013.
26. "International Association of Hydrological Sciences (IAHS)". Associations. International Union of Geodesy and Geophysics. 1 December 2008. Retrieved 8 March 2013.
27. "International Association of Hydrological Sciences". Retrieved 8 March 2013.
28. "International Commission on Statistical Hydrology". STAHY. Retrieved 8 March 2013.
29. Deutsche Hydrologische Gesellschaft, accessed 2 September 2013
30. Nordic Association for Hydrology, accessed 2 September 2013
31. "The British Hydrological Society". Retrieved 8 March 2013.
32. Гидрологическая комиссия [Hydrological Commission] (in Russian). Russian Geographical Society. Retrieved 8 March 2013.
33. «Hydroweb». The International Association for Environmental Hydrology. Retrieved 8 March 2013.
34. «International Association of Hydrogeologists». Retrieved 19 June 2014.
35. «Connected Waters Initiative (CWI)». University of New South Wales. Retrieved 8 March 2013.
36. «Integrated Water Resource Management in Australia: Case studies – Murray-Darling Basin initiative». Australian Government, Department of the Environment. Australian Government. Retrieved 19 June 2014.

37. Wetzel, R.G. (2001) Limnology: Lake and River Ecosystems, 3rd ed. Academic Press. ISBN 0-12-744760-1

CHAPTER 2

Hydrogeology

Chapter Contents

- **Introduction**
- **Hydrogeology In Relation To Other Fields**
- **Definitions And Material Properties**
- **Governing Equations**
- **Calculation Of Groundwater Flow**

Hydrogeology is a branch of geology, the study of rocks and the structures that are formed over past periods of time. The way that sediments have been deposited to create different layers of rock beneath the surface, or the way that rocks have been heated and folded over millions of years to create complex structures are the subjects of geology.

INTRODUCTION

Hydrogeology (hydro- meaning water, and -geology meaning the study of the Earth) is the area of geology that deals with the distribution and movement of groundwater in the soil and rocks of the Earth's crust (commonly in aquifers). The term geohydrology is often used interchangeably. Some make the minor distinction between a hydrologist or engineer applying themselves to geology (geohydrology), and a geologist applying themselves to hydrology (hydrogeology).

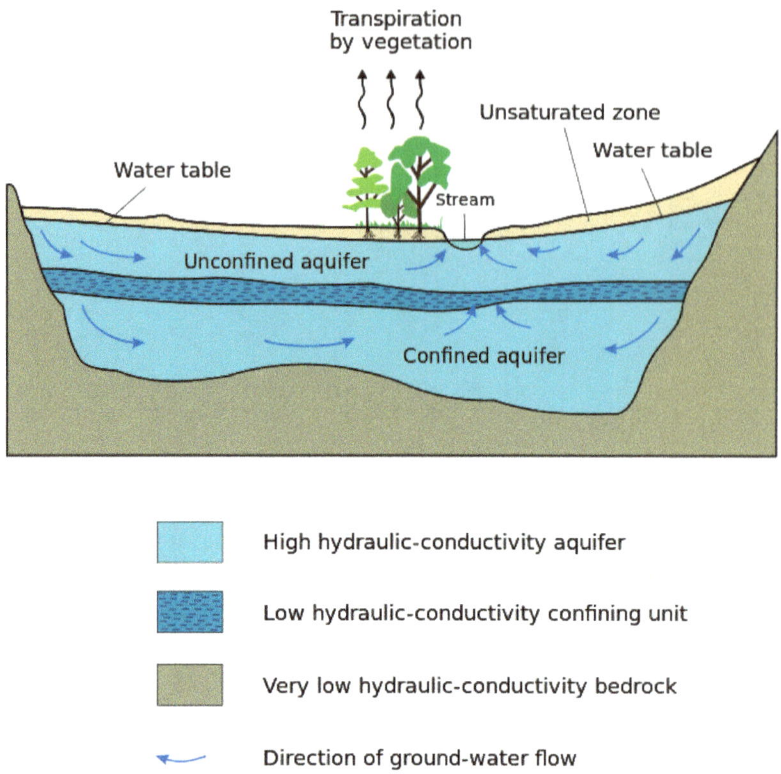

Typical aquifer cross-section

Hydrogeology is an interdisciplinary subject; it can be difficult to account fully for the chemical, physical, biological and even legal interactions between soil, water, nature and society. The study of the interaction between groundwater movement and geology can be quite complex. Groundwater does not always flow in the subsurface down-hill following the surface topography; groundwater follows pressure gradients (flow from high pressure to low) often following fractures and conduits in circuitous paths. Taking into account the interplay of the different facets of a multi-component system often requires knowledge in several diverse fields at both the experimental and theoretical levels. The following is a more traditional introduction to the methods and nomenclature of saturated subsurface hydrology, or simply the study of ground water content.

HYDROGEOLOGY IN RELATION TO OTHER FIELDS

Hydrogeology, as stated above, is a branch of the earth sciences dealing with the flow of water through aquifers and other shallow porous media (typically less than 450 m or 1,500 ft below the land surface). The very shallow flow of water in the subsurface (the upper 3 m or 10 ft) is pertinent to the fields of soil science, agriculture and civil engineering, as well as to hydrogeology. The general flow of fluids (water, hydrocarbons, geothermal fluids, etc.) in deeper formations is also a concern of geologists, geophysicists and petroleum geologists. Groundwater is a slow-moving, viscous fluid (with a Reynolds number less than unity); many of the empirically derived laws of groundwater flow can be alternately derived in fluid mechanics from the special case of Stokes flow (viscosity and pressure terms, but no inertial term).

The mathematical relationships used to describe the flow of water through porous media are the diffusion and Laplace equations, which have applications in many diverse fields. Steady groundwater flow (Laplace equation) has been simulated using electrical, elastic and heat conduction analogies. Transient groundwater flow is analogous to the diffusion of heat in a solid, therefore some solutions to hydrological problems have been adapted from heat transfer literature.

Traditionally, the movement of groundwater has been studied separately from surface water, climatology, and even the chemical and microbiological aspects of hydrogeology (the processes are uncoupled). As the field of hydrogeology matures, the strong interactions between groundwater, surface water, water chemistry, soil moisture and even climate are becoming more clear.

For example: aquifer drawdown or overdrafting and the pumping of fossil water may be a contributing factor to sea-level rise.

DEFINITIONS AND MATERIAL PROPERTIES

One of the main tasks a hydrogeologist typically performs is the prediction of future behavior of an aquifer system, based on analysis of past and present observations. Some hypothetical, but characteristic questions asked would be:

- Can the aquifer support another subdivision?
- Will the river dry up if the farmer doubles his irrigation?
- Did the chemicals from the dry cleaning facility travel through the aquifer to my well and make me sick?
- Will the plume of effluent leaving my neighbor's septic system flow to my drinking water well?

Most of these questions can be addressed through simulation of the hydrologic system (using numerical models or analytic equations). Accurate simulation of the aquifer system requires knowledge of the aquifer properties and boundary conditions. Therefore, a common task of the hydrogeologist is determining aquifer properties using aquifer tests.

In order to further characterize aquifers and aquitards some primary and derived physical properties are introduced below. Aquifers are broadly classified as being either confined or unconfined (water table aquifers), and either saturated or unsaturated; the type of aquifer affects what properties control the flow of water in that medium (e.g., the release of water from storage for confined aquifers is related to the storativity, while it is related to the specific yield for unconfined aquifers).

Hydraulic Head

Differences in hydraulic head (h) cause water to move from one place to another; water flows from locations of high h to locations of low h. Hydraulic head is composed of pressure head (ψ) and elevation head (z). The head gradient is the change in hydraulic head per length of flowpath, and appears in Darcy's law as being proportional to the discharge.

Hydraulic head is a directly measurable property that can take on any value (because of the arbitrary datum involved in the z term); ψ can be measured with a pressure transducer (this value can be negative, e.g., suction, but is positive in saturated aquifers), and z can be measured relative to a surveyed datum (typically the top of the well casing). Commonly, in wells tapping unconfined aquifers the water level in a well is used as a proxy for hydraulic head, assuming there is no vertical gradient of pressure. Often only changes in hydraulic head through time are needed, so the constant elevation head term can be left out ($\Delta h = \Delta\psi$).

A record of hydraulic head through time at a well is a hydrograph or, the changes in hydraulic head recorded during the pumping of a well in a test are called drawdown.

Porosity

Porosity (n) is a directly measurable aquifer property; it is a fraction between 0 and 1 indicating the amount of pore space between unconsolidated soil particles or within a fractured rock. Typically, the majority of groundwater (and anything dissolved in it) moves through the porosity available to flow (sometimes called effective porosity). **Permeability** is an expression of the connectedness of the pores. For instance, an unfractured rock unit may have a high porosity (it has lots of holes between its constituent grains), but a low permeability (none of the pores are connected). An example of this phenomenon is pumice, which, when in its unfractured state, can make a poor aquifer.

Porosity does not directly affect the distribution of hydraulic head in an aquifer, but it has a very strong effect on the migration of dissolved contaminants, since it affects groundwater flow velocities through an inversely proportional relationship.

Water Content

Water content (θ) is also a directly measurable property; it is the fraction of the total rock which is filled with liquid water. This is also a fraction between 0 and 1, but it must also be less than or equal to the total porosity.

The water content is very important in vadose zone hydrology, where the hydraulic conductivity is a strongly nonlinear function of water content; this complicates the solution of the unsaturated groundwater flow equation.

Hydraulic Conductivity

Hydraulic conductivity (K) and transmissivity (T) are indirect aquifer properties (they cannot be measured directly). T is the K integrated over the vertical thickness (b) of the aquifer (T=Kb when K is constant over the entire thickness). These properties are measures of an aquifer's ability to transmit water. Intrinsic permeability (κ) is a secondary medium property which does not depend on the viscosity and density of the fluid (K and T are specific to water); it is used more in the petroleum industry.

Specific Storage and Specific Yield

Specific storage (S_s) and its depth-integrated equivalent, storativity ($S=S_sb$), are indirect aquifer properties (they cannot be measured directly); they indicate the amount of groundwater released from storage due to a unit depressurization of a confined aquifer. They are fractions between 0 and 1.

Specific yield (S_y) is also a ratio between 0 and 1 ($S_y \leq$ porosity) and indicates the amount of water released due to drainage from lowering the water table in an unconfined aquifer. The value for specific yield is less than the value for porosity because some water will remain in the medium even after drainage due to intermolecular forces. Often the porosity or effective porosity is used as an upper bound to the specific yield. Typically S_y is orders of magnitude larger than S_s.

Contaminant Transport Properties

Often we are interested in how the moving groundwater will transport dissolved contaminants around (the sub-field of contaminant hydrogeology). The contaminants can be man-made (e.g., petroleum products, nitrate, Chromium or radionuclides) or naturally occurring (e.g., arsenic, salinity). Besides needing to understand where the groundwater is flowing, based on the other hydrologic properties discussed above, there are additional aquifer properties which affect how dissolved contaminants move with groundwater.

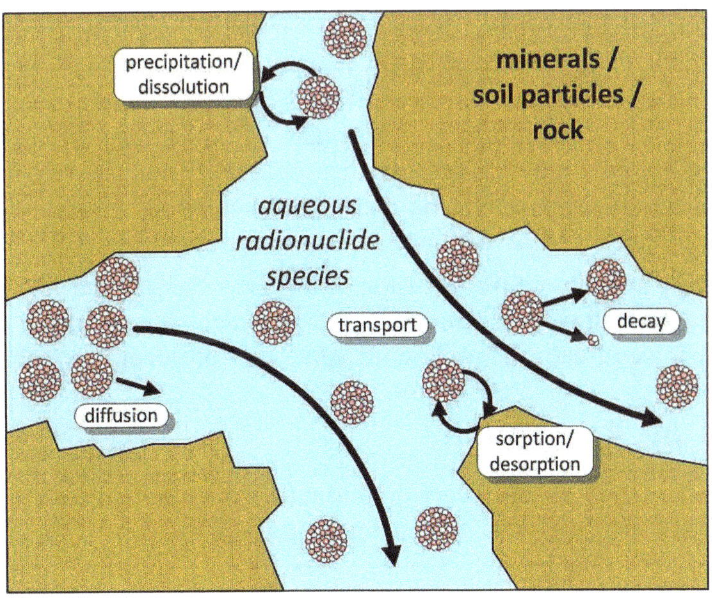

Transport and fate of contaminants in groundwater

Hydrodynamic Dispersion

Hydrodynamic dispersivity (α_L, α_T) is an empirical factor which quantifies how much contaminants stray away from the path of the groundwater which is carrying it. Some of the contaminants will be "behind" or "ahead" the mean groundwater, giving rise to a longitudinal dispersivity (α_L), and some will be "to the sides of" the pure advective groundwater flow, leading to a transverse dispersivity (α_T). Dispersion in groundwater arises because each water "particle", passing beyond a soil particle, must choose where to go, whether left or right or up or down, so that the water "particles" (and their solute) are gradually spread in all directions around the mean path. This is the "microscopic" mechanism, on the scale of soil particles. More important, on long distances, can be the macroscopic inhomogeneities of the aquifer, which can have regions of larger or smaller permeability, so that some water can find a preferential path in one direction, some other in a different direction, so that the contaminant can be spread in a completely irregular way, like in a (three-dimensional) delta of a river.

Dispersivity is actually a factor which represents our lack of information about the system we are simulating. There are many small details about the aquifer which are being averaged when using a macroscopic approach (e.g., tiny beds of gravel and clay in sand aquifers), they manifest themselves as an apparent dispersivity. Because of this, α is often claimed to be dependent on the length scale of the problem — the dispersivity found for transport through 1 m^3 of aquifer is different from that for transport through 1 cm^3 of the same aquifer material.

Molecular Diffusion

Diffusion is a fundamental physical phenomenon, which Einstein characterized as Brownian motion, that describes the random thermal movement of molecules and small particles in gases and liquids. It is an important phenomenon for small distances (it is essential for the achievement of thermodynamic equilibria), but, as the time necessary to cover a distance by diffusion is proportional to the square of the distance itself, it is ineffective for spreading a solute over macroscopic distances. The diffusion coefficient, D, is typically quite small, and its effect can often be considered negligible (unless groundwater flow velocities are extremely low, as they are in clay aquitards).

It is important not to confuse diffusion with dispersion, as the former is a physical phenomenon and the latter is an empirical factor which is cast into a similar form as diffusion, because we already know how to solve that problem.

Retardation by Adsorption

The retardation factor is another very important feature that make the motion of the contaminant to deviate from the average groundwater motion. It is analogous to the retardation factor of chromatography. Unlike diffusion and dispersion, which simply spread the contaminant, the retardation factor changes its global average velocity, so that it can be much slower than that of water. This is due to a chemico-physical effect: the adsorption to the soil, which holds the contaminant back and does not allow it to progress until the quantity corresponding to the chemical adsorption equilibrium has been adsorbed. This effect is particularly important for less soluble contaminants, which thus can move even hundreds or thousands times slower than water. The effect of this phenomenon is that only more soluble species can cover long distances. The retardation factor depends on the chemical nature of both the contaminant and the aquifer.

GOVERNING EQUATIONS

Darcy's Law

Darcy's law is a constitutive equation, empirically derived by Henry Darcy in 1856, which states that the amount of groundwater discharging through a given portion of aquifer is proportional to the cross-sectional area of flow, the hydraulic gradient, and the hydraulic conductivity.

Groundwater Flow Equation

The groundwater flow equation, in its most general form, describes the movement of groundwater in a porous medium (aquifers and aquitards). It is known in mathematics as the diffusion equation, and has many analogs in other fields. Many solutions for groundwater flow problems were borrowed or adapted from existing heat transfer solutions.

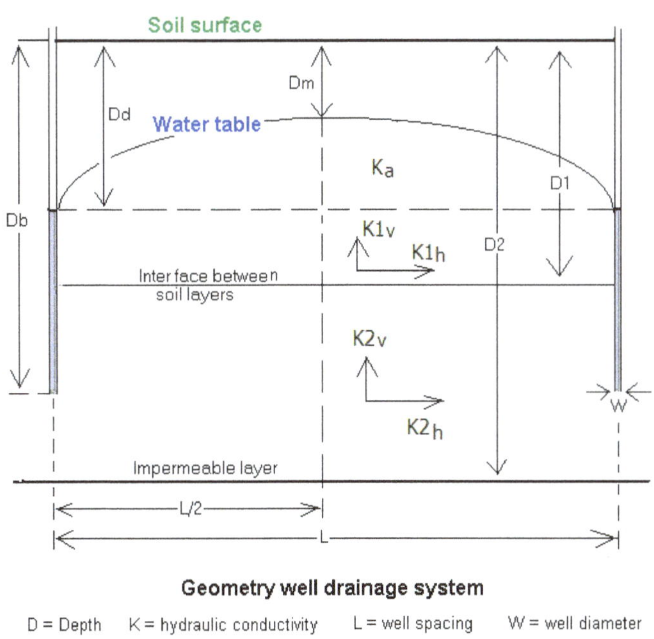

Geometry of a partially penetrating well drainage system in an anisotropic layered aquifer

It is often derived from a physical basis using Darcy's law and a conservation of mass for a small control volume. The equation is often used to predict flow to wells, which have radial symmetry, so the flow equation is commonly solved in polar or cylindrical coordinates.

The Theis equation is one of the most commonly used and fundamental solutions to the groundwater flow equation; it can be used to predict the transient evolution of head due to the effects of pumping one or a number of pumping wells.

The Thiem equation is a solution to the steady state groundwater flow equation (Laplace's Equation) for flow to a well. Unless there are large sources of water nearby (a river or lake), true steady-state is rarely achieved in reality.

Both above equations are used in aquifer tests (pump tests).

The Hooghoudt equation is a groundwater flow equation applied to subsurface drainage by pipes, tile drains or ditches. An alternative subsurface drainage method is drainage by wells for which groundwater flow equations are also available.

CALCULATION OF GROUNDWATER FLOW

To use the groundwater flow equation to estimate the distribution of hydraulic heads, or the direction and rate of groundwater flow, this partial differential equation (PDE) must be solved. The most common means of analytically solving the diffusion equation in the hydrogeology literature are:

- Laplace, Hankel and Fourier transforms (to reduce the number of dimensions of the PDE),
- similarity transform (also called the Boltzmann transform) is commonly how the Theis solution is derived,
- separation of variables, which is more useful for non-Cartesian coordinates, and
- Green's functions, which is another common method for deriving the Theis solution — from the fundamental solution to the diffusion equation in free space.

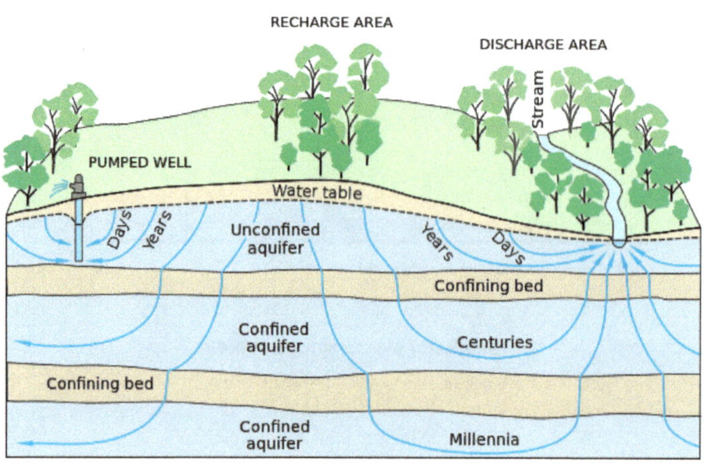

Relative groundwater travel times.

No matter which method we use to solve the groundwater flow equation, we need both initial conditions (heads at time (t) = 0) and boundary conditions (representing either the physical boundaries of the domain, or an approximation of the domain beyond that point). Often the initial conditions are supplied to a transient simulation, by a corresponding steady-state simulation (where the time derivative in the groundwater flow equation is set equal to 0).

There are two broad categories of how the (PDE) would be solved; either analytical methods, numerical methods, or something possibly in between. Typically, analytic methods solve the groundwater flow equation under a simplified set of conditions exactly, while numerical methods solve it under more general conditions to an approximation.

Analytic Methods

Analytic methods typically use the structure of mathematics to arrive at a simple, elegant solution, but the required derivation for all but the simplest domain geometries can be quite complex (involving non-standard coordinates, conformal mapping, etc.). Analytic solutions typically are also simply an equation that can give a quick answer based on a few basic parameters. The Theis equation is a very simple (yet still very useful) analytic solution to the groundwater flow equation, typically used to analyze the results of an aquifer test or slug test.

Numerical Methods

The topic of numerical methods is quite large, obviously being of use to most fields of engineering and science in general. Numerical methods have been around much longer than computers have (In the 1920s Richardson developed some of the finite difference schemes still in use today, but they were calculated by hand, using paper and pencil, by human "calculators"), but they have become very important through the availability of fast and cheap personal computers.

There are two broad categories of numerical methods: gridded or discretized methods and non-gridded or mesh-free methods. In the common finite difference method and finite element method (FEM) the domain is completely gridded ("cut" into a grid or mesh of small elements). The analytic element method (AEM) and the boundary integral equation method (BIEM — sometimes also called BEM, or Boundary Element Method) are only discretized at boundaries or along flow elements (line sinks, area sources, etc.), the majority of the domain is mesh-free.

General Properties of Gridded Methods

Gridded Methods like finite difference and finite element methods solve the groundwater flow equation by breaking the problem area (domain) into many small elements (squares, rectangles, triangles, blocks, tetrahedra, etc.) and solving the flow equation for each element (all material properties are assumed constant or possibly linearly variable within an element), then linking together all the elements using conservation of mass across the boundaries between the elements (similar to the divergence theorem). This results in a system which overall approximates the groundwater flow equation, but exactly matches the boundary conditions (the head or flux is specified in the elements which intersect the boundaries).

Finite differences are a way of representing continuous differential operators using discrete intervals (Δx and Δt), and the finite difference methods are based on these (they are derived from a Taylor series). For example, the first-order time derivative is often approximated using the following forward finite difference, where the subscripts indicate a discrete time location,

$$\frac{\partial h}{\partial t} = h'(t_i) \approx \frac{h_i - h_{i-1}}{\Delta t}.$$

The forward finite difference approximation is unconditionally stable, but leads to an implicit set of equations (that must be solved using matrix methods, e.g. LU or Cholesky decomposition). The similar backwards difference is only conditionally stable, but it is explicit and can be used to "march" forward in the time direction, solving one grid node at a time (or possibly in parallel, since one node depends only on its immediate neighbors). Rather than the finite difference method, sometimes the Galerkin FEM approximation is used in space (this is different from the type of FEM often used in structural engineering) with finite differences still used in time.

Application of Finite Difference Models

MODFLOW is a well-known example of a general finite difference groundwater flow model. It is developed by the US Geological Survey as a modular and extensible simulation tool for modeling groundwater flow. It is free software developed, documented and distributed by the USGS. Many commercial products have grown up around it, providing graphical user interfaces to its input file based interface, and typically incorporating pre- and post-processing of user data. Many other models have been developed to work with MODFLOW input and output, making linked models which simulate several hydrologic processes possible (flow and transport models, surface water and groundwater models and chemical reaction models), because of the simple, well documented nature of MODFLOW.

Application of Finite Element Models

Finite Element programs are more flexible in design (triangular elements vs. the block elements most finite difference models use) and there are some programs available (SUTRA, a 2D or 3D density-dependent flow model by the USGS; Hydrus, a commercial unsaturated flow model; FEFLOW, a commercial modelling environment for subsurface flow, solute and heat transport processes; OpenGeoSys, a scientific open-source project for thermo-hydro-mechanical-chemical (THMC) processes in porous and fractured media; COMSOL Multiphysics (FEMLAB) a commercial general modelling environment), FEATool Multiphysics, an easy to use Matlab simulation toolbox, and Integrated Water Flow Model (IWFM), but they are still not as popular in with practicing hydrogeologists as MODFLOW is. Finite element models are more popular in university and laboratory environments, where specialized models solve non-standard forms of the flow equation (unsaturated flow, density dependent flow, coupled heat and groundwater flow, etc.)

Application of Finite Volume Models

The finite volume method is a method for representing and evaluating partial differential equations as algebraic equations. Similar to the finite difference method, values are calculated at discrete places on a meshed geometry. "Finite volume" refers to the small volume surrounding each node point on a mesh. In the finite volume method, volume integrals in a partial differential equation that contain a divergence term are converted to surface integrals, using the divergence theorem. These terms are then evaluated as fluxes at the surfaces of each finite volume. Because the flux entering a given volume is identical to that leaving the adjacent volume, these methods are conservative. Another advantage of the finite volume method is that it is easily formulated to allow for unstructured meshes. The method is used in many computational fluid dynamics packages.

PORFLOW software package is a comprehensive mathematical model for simulation of Ground Water Flow and Nuclear Waste Management developed by Analytic & Computational Research, Inc., ACRi.

The FEHM software package is available free from Los Alamos National Laboratory. This versatile porous flow simulator includes capabilities to model multiphase, thermal, stress, and multicomponent reactive chemistry. Current work using this code includes simulation of methane hydrate formation, CO_2 sequestration, oil shale extraction, migration of both nuclear and chemical contaminants, environmental isotope migration in the unsaturated zone, and karst formation.

Other Methods

These include mesh-free methods like the Analytic Element Method (AEM) and the Boundary Element Method (BEM), which are closer to analytic solutions, but they do approximate the groundwater flow equation in some way. The BEM and AEM exactly solve the groundwater flow equation (perfect mass balance), while approximating the boundary conditions. These methods are more exact and can be much more elegant solutions (like analytic methods are), but have not seen as widespread use outside academic and research groups yet.

REFERENCES

1. "Rising sea levels attributed to global groundwater extraction". University of Utrecht. Retrieved February 8, 2011.
2. http://www.cof.orst.edu/cof/fe/watershd/fe537/labs_2007/gelhar_etal_reviewfieldScaleDispersion_WRR1992.pdf
3. The energy balance of groundwater flow applied to subsurface drainage in anisotropic soils by pipes or ditches with entrance resistance. International Institute for Land Reclamation and Improvement (ILRI), Wageningen, The Netherlands.

On line : . Paper based on: R.J. Oosterbaan, J. Boonstra and K.V.G.K. Rao, 1996, "The energy balance of groundwater flow". Published in V.P.Singh and B.Kumar (eds.), Subsurface-Water Hydrology, p. 153-160, Vol.2 of Proceedings of the International Conference on Hydrology and Water Resources, New Delhi, India, 1993. Kluwer Academic Publishers, Dordrecht, The Netherlands. ISBN 978-0-7923-3651-8 . On line : . The corresponding free computer program EnDrain can be downloaded from web page : , or from :

4. ILRI, 2000, Subsurface drainage by (tube)wells: Well spacing equations for fully and partially penetrating wells in uniform or layered aquifers with or without anisotropy and entrance resistance, 9 pp. Principles used in the "WellDrain" model. International Institute for Land Reclamation and Improvement (ILRI), Wageningen, The Netherlands. On line: . Free download "WellDrain" software from web page : , or from :
5. "OpenGeoSys". Helmholtz centre for environmental research. Retrieved 18 May 2012.
6. "OpenGeoSys Website". Helmholtz centre for environmental research. Retrieved 28 April 2014.
7. FEATool Multiphysics homepage
8. LeVeque, 2002
9. Toro, 1999

CHAPTER 3

Evapotranspiration

Chapter Contents

- **Introduction**
- **Water Cycle**
- **Estimating Evapotranspiration**
- **Potential Evapotranspiration**

The term evapotranspiration combines two words: evaporation of water from the soil, and transpiration of water from plants into the air. Evapotranspiration means the total loss of water from a crop into the air. Water evaporates from any moist surface into the air unless the air is saturated. Water surfaces in contact with air, such as lakes, plant leaves, and moist soils, all evaporate water.

INTRODUCTION

Evapotranspiration (ET) is the sum of evaporation and plant transpiration from the Earth's land and ocean surface to the atmosphere. Evaporation accounts for the movement of water to the air from sources such as the soil, canopy interception, and waterbodies. Transpiration accounts for the movement of water within a plant and the subsequent loss of water as vapor through stomata in its leaves. Evapotranspiration is an important part of the water cycle. An element (such as a tree) that contributes to evapotranspiration can be called an evapotranspirator.

Reference evapotranspiration (ET_0), sometimes incorrectly referred to as potential ET, is a representation of the environmental demand for evapotranspiration and represents the evapotranspiration rate of a short green crop (grass), completely shading the ground, of uniform height and with adequate water status in the soil profile. It is a reflection of the energy available to evaporate water, and of the wind available to transport the water vapour from the ground up into the lower atmosphere. Actual evapotranspiration is said to equal reference evapotranspiration when there is ample water. Some US states utilize a full cover alfalfa reference crop that is 0.5 m in height, rather than the short green grass reference, due to the higher value of ET from the alfalfa reference.

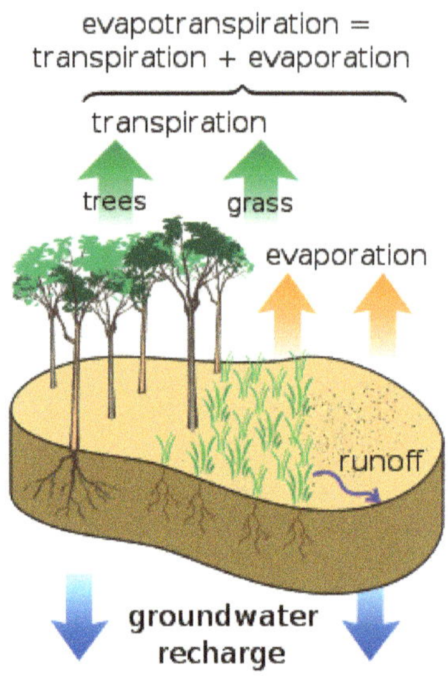

Water cycle of the Earth's surface, showing the individual components of transpiration and evaporation that make up evapotranspiration. Other closely related processes shown are runoff and groundwater recharge.

WATER CYCLE

Evapotranspiration is a significant water loss from drainage basins. Types of vegetation and land use significantly affect evapotranspiration, and therefore the amount of water leaving a drainage basin. Because water transpired through leaves comes from the roots, plants with deep reaching roots can more constantly transpire water. Herbaceous plants generally transpire less than woody plants because they usually have less extensive foliage. Conifer forests tend to have higher rates of evapotranspiration than deciduous forests, particularly in the dormant and early spring seasons. This is primarily due to the enhanced amount of precipitation intercepted and evaporated by conifer foliage during these periods. Factors that affect evapotranspiration include the plant's growth stage or level of maturity, percentage of soil cover, solar radiation, humidity, temperature, and wind. Isotope measurements indicate transpiration is the larger component of evapotranspiration.

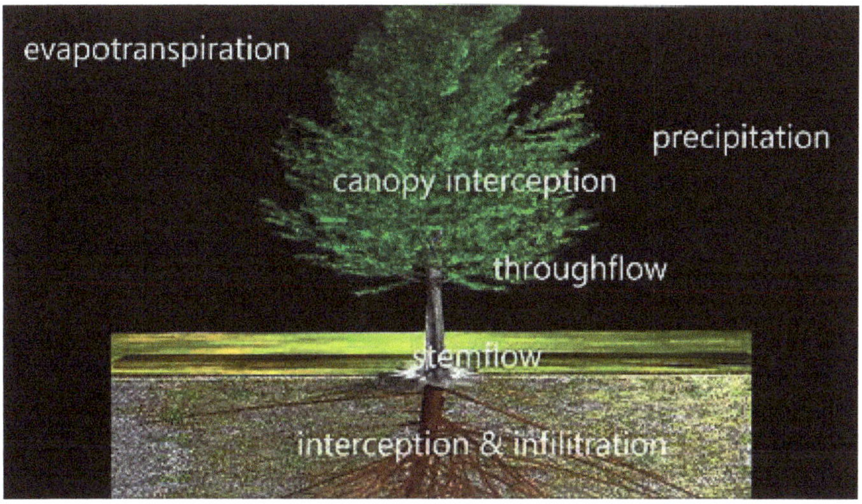

Through evapotranspiration, forests reduce water yield, except in unique ecosystems called cloud forests. Trees in cloud forests collect the liquid water in fog or low clouds onto their surface, which drips down to the ground. These trees still contribute to evapotranspiration, but often collect more water than they evaporate or transpire.

In areas that are not irrigated, actual evapotranspiration is usually no greater than precipitation, with some buffer in time depending on the soil's ability to hold water. It will usually be less because some water will be lost due to percolation or surface runoff. An exception is areas with high water tables, where capillary action can cause water from the groundwater to rise through the soil matrix to the surface. If potential evapotranspiration is greater than actual precipitation, then soil will dry out, unless irrigation is used.

Evapotranspiration can never be greater than PET, but can be lower if there is not enough water to be evaporated or plants are unable to transpire readily.

ESTIMATING EVAPOTRANSPIRATION

Evapotranspiration can be measured or estimated using several methods.

Indirect Methods

Pan evaporation data can be used to estimate lake evaporation, but transpiration and evaporation of intercepted rain on vegetation are unknown. There are three general approaches to estimate evapotranspiration indirectly.

Catchment Water Balance

Evapotranspiration may be estimated by creating an equation of the water balance of a drainage basin. The equation balances the change in water stored within the basin (S) with inputs and outgoes:

$$\Delta S = P - ET - Q - D$$

The input is precipitation (P) and the outgoes are evapotranspiration (which is to be estimated), streamflow (Q), and groundwater recharge (D). If the change in storage, precipitation, streamflow, and groundwater recharge are all estimated, the missing flux, ET, can be estimated by rearranging the above equation as follows:

$$ET = P - \Delta S - Q - D$$

Evapotranspiration of Urban Vegetation

The water requirement of urban landscapes, particularly urban parklands, is of growing concern. The estimation of evapotranspiration (ET) and subsequently plant water requirements in urban vegetation needs to consider the heterogeneity of plants, soils, water, and climate characteristics. In a research in South Australia, two practical ET estimation approaches are compared to a detailed Soil Water Balance (SWB) analysis over a one-year period. One approach is the Water Use Classification of Landscape Plants (WUCOLS) method, which is based on expert opinion on the water needs of different classes of landscape plants. The other is a remote sensing approach based on the Enhanced Vegetation Index (EVI) from Moderate Resolution Imaging Spectroradiometer (MODIS) sensors on the Terra satellite. Both methods require knowledge of reference ET calculated from meteorological data. More information is available in:

- Comparing Three Approaches of Evapotranspiration Estimation in Mixed Urban Vegetation: Field-Based, Remote Sensing-Based and Observational-Based Methods
- Water requirements of urban landscape plants: a comparison of three factor-based approaches
- High Spatial Resolution WorldView-2 Imagery for Mapping NDVI and Its Relationship to Temporal Urban Landscape Evapotranspiration Factors*

Hydrometeorological Equations

The most general and widely used equation for calculating reference ET is the Penman equation. The Penman-Monteith variation is recommended by the Food and Agriculture Organization and the American Society of Civil Engineers. The simpler Blaney-Criddle equation was popular in the Western United States for many years but it is not as accurate in regions with higher humidities. Other solutions used includes Makkink, which is simple but must be calibrated to a specific location, and Hargreaves. To convert the reference evapotranspiration to actual crop evapotranspiration, a crop coefficient and a stress coefficient must be used. Crop coefficients referred to in many hydrological models are themselves during periods for which the model is used. This is because crops are seasonal, perennial plants mature over multiple seasons, and stress responses can significantly depend upon many aspects of plant condition.

Energy Balance

A third methodology to estimate the actual evapotranspiration is the use of the energy balance.

$$\lambda E = R_n - G - H$$

where λE is the energy needed to change the phase of water from liquid to gas, R_n is the net radiation, G is the soil heat flux and H is the sensible heat flux. Using instruments like a scintillometer, soil heat flux plates or radiation meters, the components of the energy balance can be calculated and the energy available for actual evapotranspiration can be solved.

The SEBAL and METRIC algorithms solve the energy balance at the earth's surface using satellite imagery. This allows for both actual and potential evapotranspiration to be calculated on a pixel-by-pixel basis. Evapotranspiration is a key indicator for water management and irrigation performance. SEBAL and METRIC can map these key indicators in time and space, for days, weeks or years.

Experimental Methods for Measuring Evapotranspiration

One method for measuring evapotranspiration is with a weighing lysimeter. The weight of a soil column is measured continuously and the change in storage of water in the soil is modeled by the change in weight. The change in weight is converted to units of length using the surface area of the weighing lysimeter and the unit weight of water. evapotranspiration is computed as the change in weight plus rainfall minus percolation.

Remote Sensing

In recent decades, estimating evapotranspiration has been improved by advances in remote sensing, particularly in agricultural studies. However, quantifying evapotranspiration from mixed vegetation environs, particularly urban parklands, is still challenging because of the heterogeneity of plant species, canopy covers and microclimates and because the methodology is costly. Different remote sensing-based approaches for estimating evapotranspiration have various advantages and disadvantages.

Eddy Covariance

The eddy covariance (also known as eddy correlation and eddy flux) technique is a key atmospheric measurement technique to measure and calculate vertical turbulent fluxes within atmospheric boundary layers. The method analyzes high-frequency wind and scalar atmospheric data series, and yields values of fluxes of these properties. It is a statistical method used in meteorology and other applications (micrometeorology, oceanography, hydrology, agricultural sciences, industrial and regulatory applications, etc.) to determine exchange rates of trace gases over natural ecosystems and agricultural fields, and to quantify gas emissions rates from other land and water areas. It is frequently used to estimate momentum, heat, water vapour, carbon dioxide and methane fluxes.

The technique is also used extensively for verification and tuning of global climate models, mesoscale and weather models, complex biogeochemical and ecological models, and remote sensing estimates from satellites and aircraft. The technique is mathematically complex, and requires significant care in setting up and processing data. To date, there is no uniform terminology or a single methodology for the Eddy Covariance technique, but much effort is being made by flux measurement networks (e.g., FluxNet, Ameriflux, ICOS, CarboEurope, Fluxnet Canada, OzFlux, NEON, and iLEAPS) to unify the various approaches.

Eddy covariance system consisting of an ultrasonic anemometer and infrared gas analyser (IRGA).

An eddy correlation instrument measuring oxygen fluxes in benthic environments.

The technique has additionally proven applicable under water to the benthic zone for measuring oxygen fluxes between seafloor and overlying water. In these environments, the technique is generally known as the eddy correlation technique, or just eddy correlation. Oxygen fluxes are extracted from raw measurements largely following the same principles as used in the atmosphere, and they are typically used as a proxy for carbon exchange, which is important for local and global carbon budgets. For most benthic ecosystems, eddy correlation is the most accurate technique for measuring *in-situ* fluxes. The technique's development and its applications under water remains a fruitful area of research.

General Principles

Representation of the Air Flow in the Atmospheric Boundary Layer

Air flow can be imagined as a horizontal flow of numerous rotating eddies, that is, turbulent vortices of various sizes, with each eddy having horizontal and vertical components. The situation looks chaotic, but vertical movement of the components can be measured from the tower.

Physical Meaning of the Eddy Covariance Method

At one physical point on the tower, at Time1, Eddy1 moves parcel of air c1 down at the speed w1. Then, at Time2, Eddy2 moves parcel c2 up at the speed w2. Each parcel has gas concentration, pressure, temperature, and humidity. If these factors, along with the speed are known, we can determine the flux. For example, if one knew how many molecules of water went down with eddies at Time 1, and how many molecules went up with eddies at Time2, at the same point, one could calculate the vertical flux of water at this point over this time. So, vertical flux can be presented as a covariance of the vertical wind velocity and the concentration of the entity of interest.

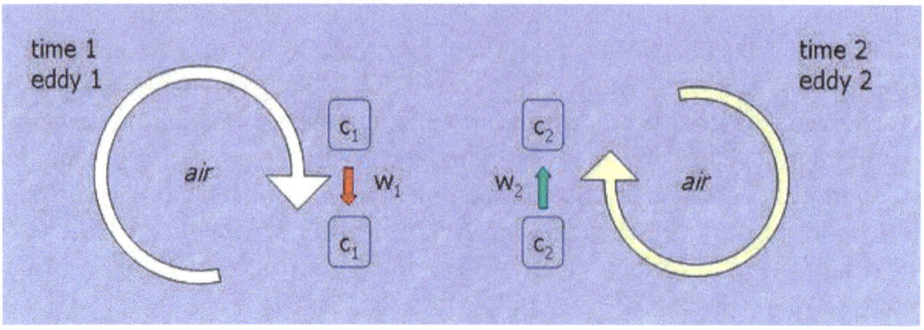

Mathematical Foundation

In mathematical terms, "eddy flux" is computed as a covariance between instantaneous deviation in vertical wind speed (w') from the mean value (w-overbar) and instantaneous deviation in gas concentration, mixing ratio (s'), from its mean value (s-overbar), multiplied by mean air density (ρa). Several mathematical operations and assumptions, including Reynolds decomposition, are involved in getting from physically complete equations of the turbulent flow to practical equations for computing "eddy flux", as shown below.

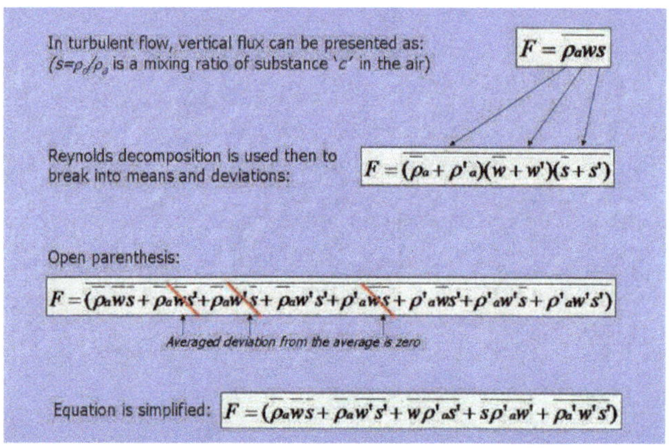

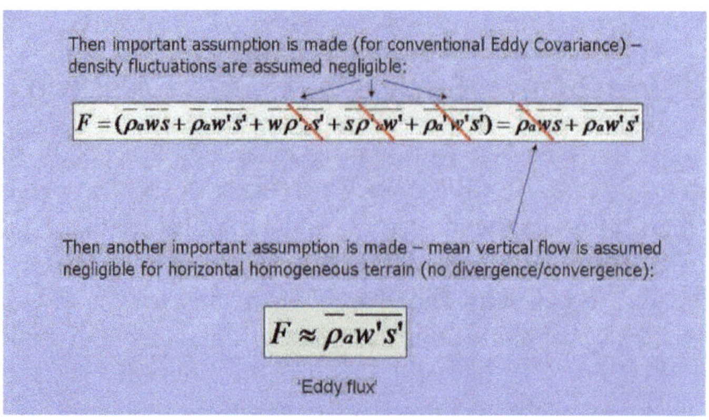

Major Assumptions

- Measurements at a point can represent an upwind area
- Measurements are done inside the boundary layer of interest
- Fetch/flux footprint is adequate – fluxes are measured only at area of interest
- Flux is fully turbulent – most of the net vertical transfer is done by eddies
- Terrain is horizontal and uniformed: average of fluctuations is zero; density fluctuations negligible; flow convergence & divergence negligible
- Instruments can detect very small changes at high frequency, ranging from minimum of 5 Hz and to 40 Hz for tower-based measurements

Eddy Covariance Processing Software

Currently (2011) there are many software programs to process eddy covariance data and derive quantities such as heat, momentum, and gas fluxes. The programs range significantly in complexity, flexibility, number of allowed instruments and variables, help system and user support. Some programs are open-source software, while others are closed-source or proprietary.

Examples include free fully supported and documented open-source software such as EddyPro; free unsupported open-source programs such as ECO_2S and ECpack; free closed-source packages such as EdiRe, TK3, Alteddy, and EddySoft. Eddy covariance processing algorithms can also be implemented in commercial software such as Matlab.

Summary

In summation, the 3D wind and another variable (usually gas concentration, temperature or momentum) are decomposed into mean and fluctuating components. The covariance is calculated between the fluctuating component of the vertical wind and the fluctuating component of gas concentration. The measured flux is proportional to the covariance.

The area from which the detected eddies originate is described probabilistically and called a flux footprint. The flux footprint area is dynamic in size and shape, changing with wind direction, thermal stability and measurements height, and has a gradual border.

The effect of sensor separation, finite sampling length, sonic path averaging, as well as other instrumental limitations, affect frequency response of the measurement system and may need a co-spectral correction, especially noticeable with closed-path instruments and at low heights below 1-1.5 m.

Uses of Eddy Covariance Method

Common uses:

- Greenhouse gas emissions
- Carbon dioxide emission monitoring

- Methane emissions monitoring
- Water loss, evapotranspiration
- Instantaneous water use efficiency
- Instantaneous radiation use efficiency

Novel uses:

- Precision irrigation, precision agriculture
- Carbon Sequestration and Capture monitoring
- Landfill gas emissions into the atmosphere
- Emissions of gases displaced by hydraulic fracturing into the atmosphere
- Gas leak detection and location
- Methane emission from permafrost regions
- Biogenic VOCs emission
- Reactive trace gas exchange flux measurement

Urban Landscape Plants

Methods for measuring evapotranspiration can be adapted to an urban setting to estimate the water requirements of urban landscape vegetation.

POTENTIAL EVAPOTRANSPIRATION

Potential evapotranspiration (PET) is the amount of water that would be evaporated and transpired if there were sufficient water available. This demand incorporates the energy available for evaporation and the ability of the lower atmosphere to transport evaporated moisture away from the land surface. Potential evapotranspiration is higher in the summer, on less cloudy days, and closer to the equator, because of the higher levels of solar radiation that provides the energy for evaporation. Potential evapotranspiration is also higher on windy days because the evaporated moisture can be quickly moved from the ground or plant surface, allowing more evaporation to fill its place.

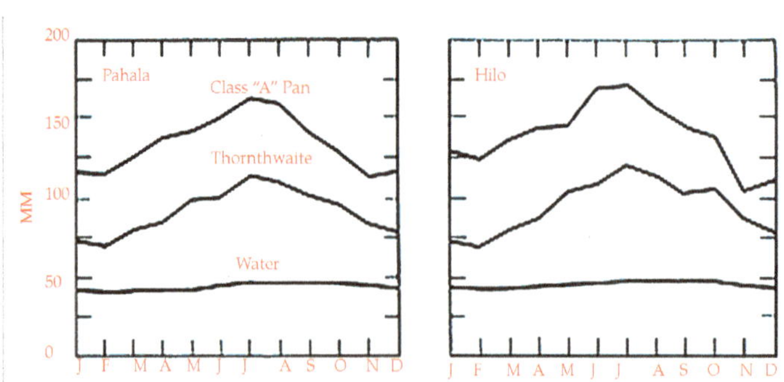

Monthly estimated potential evapotranspiration and measured pan evaporation for two locations in Hawaii, Hilo and Pahala.

Potential evapotranspiration is expressed in terms of a depth of water, and can be graphed during the year (see figure).

Potential evapotranspiration is usually measured indirectly, from other climatic factors, but also depends on the surface type, such as free water (for lakes and oceans), the soil type for bare soil, and the vegetation. Often a value for the potential evapotranspiration is calculated at a nearby climate station on a reference surface, conventionally short grass. This value is called the reference evapotranspiration, and can be converted to a potential evapotranspiration by multiplying with a surface coefficient. In agriculture, this is called a crop coefficient. The difference between potential evapotranspiration and precipitation is used in irrigation scheduling.

Average annual potential evapotranspiration is often compared to average annual precipitation, P. The ratio of the two, P/PET, is the aridity index.

REFERENCES

1. http://www.oslpr.org/download/en/2000/0031.pdf
2. http://extension.uidaho.edu/kimberly/files/2013/02/ascestzdetmain2005.pdf
3. Swank, W., and Douglass, J. 1974, Science. 185(4154) 857-859
4. Jasechko, Scott; Sharp, Zachary D.; Gibson, John J.; Birks, S. Jean; Yi, Yi; Fawcett, Peter J. (3 April 2013). "Terrestrial water fluxes dominated by transpiration". Nature. 496 (7445): 347–50. doi:10.1038/nature11983. PMID 23552893. Retrieved 4 April 2013.
5. Allen, R.G.; Pereira, L.S.; Raes, D.; Smith, M. (1998). Crop Evapotranspiration: Guidelines for Computing Crop Water Requirements. FAO Irrigation and drainage paper 56. Rome, Italy: Food and Agriculture Organization of the United Nations. ISBN 92-5-104219-5. Retrieved 2011-06-08.
6. "SEBAL_ WaterWatch".
7. Nouri, Hamideh; Beecham, Simon; Anderson, Sharoyn; Hassanli, Morad; Kazemi, Fatemeh (13 May 2014). "Remote sensing techniques for predicting evapotranspiration from mixed vegetated surfaces". Urban Water J. doi:10.1080/1573062X.2014.900092.
8. Nouri, Hamideh; Beecham, Simon; Kazemi, Fatemeh; Hassanli, Ali Morad (2013). "A review of ET measurement techniques for estimating the water requirements of urban landscape vegetation". Urban Water J. 10 (4): 247–259. doi:10.1080/1573062X.2012.726360.
9. Baldocchi, D., B. Hicks, and T. Meyers. 1988. Measuring biosphere-atmosphere exchanges of biologically related gases with micrometeorological methods. Ecology 69, 1331-1340
10. Verma, S.B.: 1990, Micrometeorological methods for measuring surface fluxes of mass and energy, Remote Sensing Reviews 5(1): 99-115
11. Lee, X., W. Massman, and B. Law. 2004. Handbook of Micrometeorology. Kluwer Academic Publishers, The Netherlands, 250 pp.

12. Burba, G., 2013. Eddy Covariance Method for Scientific, Industrial, Agricultural and Regulatory Applications: a Field Book on Measuring Ecosystem Gas Exchange and Areal Emission Rates. LI-COR Biosciences, Lincoln, USA, 331 pp.
13. Aubinet, M., T. Vesala, D. Papale (Eds.), 2012. Eddy Covariance: A Practical Guide to Measurement and Data Analysis. Springer Atmospheric Sciences, Springer Verlag, 438 pp.
14. Berg, P., H. Røy, F. Janssen, V. Meyer, B. B. Jørgensen, M. Hüttel, and D. de Beer. 2003. Oxygen uptake by aquatic sediments measured with a novel non-invasive eddy correlation technique. Marine Ecology Progress Series. 261:75-83.
15. University of Virginia. Aquatic Eddy Covariance Research Lab. Retrieved: 22 June 2015.
16. The Florida State University. Eddy Correlation - Further Development and Studies of Flow and Light driven dynamics of Benthic Oxygen Exchange. Retrieved: 22 June 2015.
17. Leibniz-Institute of Freshwater Ecology and Inland Fisheries. Eddy Correlation in Natural Waters. Retrieved: 22 June 2015.
18. Max Planck Institute for Marine Microbiology. Eddy Correlation System (ECS). Retrieved: 22 June 2015.
19. Centre for Coastal Biogeochemistry Research. Eddy Correlation. Retrieved: 22 June 2015.
20. M. Mauder, T. Foken, R. Clement, J. A. Elbers, W. Eugster, T. Grunwald, B. Heusinkveld, and O. Kolle. 2007. Quality control of CarboEurope flux data – Part II: Inter-comparison of eddy-covariance software, Biogeosciences Discuss., 4, 4067–4099

CHAPTER 4

Precipitation

Chapter Contents

Precipitation is any form of liquid or solid water particles that fall from the atmosphere and reach the surface of the Earth. For the Gulf Coast area, precipitation includes drizzle, rain, hail, and on rare occasions, snow and sleet. Different seasons and geographic locations see varying amounts of precipitation in amount and intensity. Precipitation is caused when a mass of warm, moist air hits a mass of cold air. Condensation causes the moisture to form droplets that become rain or crystals that become snow or ice. When these droplets or crystals become too heavy to be suspended in the atmosphere, they fall to Earth as precipitation.

INTRODUCTION

In meteorology, precipitation is any product of the condensation of atmospheric water vapor that falls under gravity. The main forms of precipitation include drizzle, rain, sleet, snow, graupel and hail. Precipitation occurs when a portion of the atmosphere becomes saturated with water vapor, so that the water condenses and "precipitates". Thus, fog and mist are not precipitation but suspensions, because the water vapor does not condense sufficiently to precipitate. Two processes, possibly acting together, can lead to air becoming saturated: cooling the air or adding water vapor to the air. Precipitation forms as smaller droplets coalesce via collision with other rain drops or ice crystals within a cloud. Short, intense periods of rain in scattered locations are called "showers."

Moisture overriding associated with weather fronts is an overall major method of precipitation production. If enough moisture and upward motion is present, precipitation falls from convective clouds such as cumulonimbus and can organize into narrow rainbands. Where relatively warm water bodies are present, for example due to water evaporation from lakes, lake-effect snowfall becomes a concern downwind of the warm lakes within the cold cyclonic flow around the backside of extratropical cyclones. Lake-effect snowfall can be locally heavy. Thundersnow is possible within a cyclone's comma head and within lake effect precipitation bands. In mountainous areas, heavy precipitation is possible where upslope flow is maximized within windward sides of the terrain at elevation. On the leeward side of mountains, desert climates can exist due to the dry air caused by compressional heating. The movement of the monsoon trough, or intertropical convergence zone, brings rainy seasons to savannah climes.

Precipitation is a major component of the water cycle, and is responsible for depositing the fresh water on the planet. Approximately 505,000 cubic kilometres (121,000 cu mi) of water falls as precipitation each year; 398,000 cubic kilometres (95,000 cu mi) of it over the oceans and 107,000 cubic kilometres (26,000 cu mi) over land. Given the Earth's surface area, that means the globally averaged annual precipitation is 990 millimetres (39 in), but over land it is only 715 millimetres (28.1 in). Climate classification systems such as the Köppen climate classification system use average annual rainfall to help differentiate between differing climate regimes.

Precipitation may occur on other celestial bodies, e.g. when it gets cold, Mars has precipitation which most likely takes the form of ice needles, rather than rain or snow.

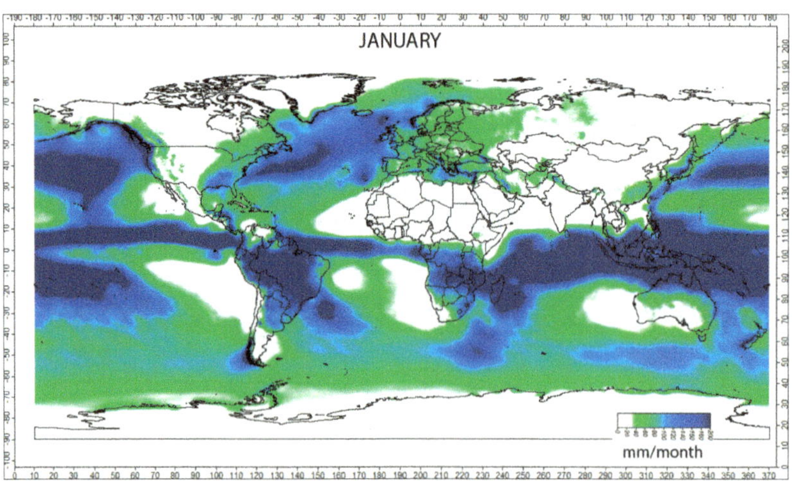

Long-term mean precipitation by month

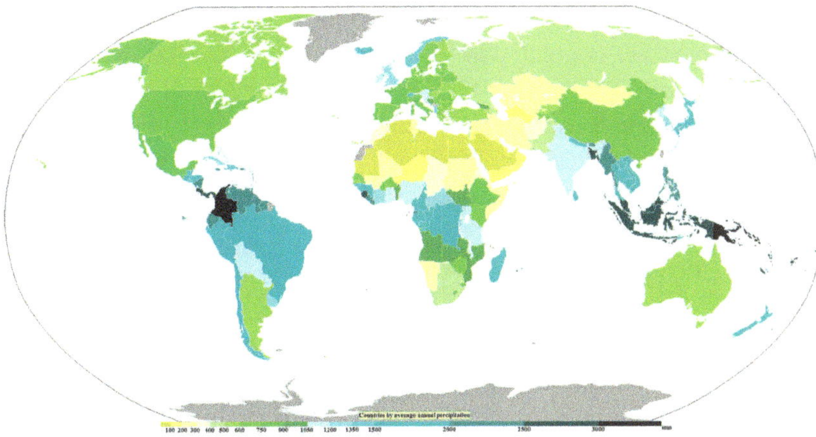

Countries by average annual precipitation

TYPES

Precipitation is a major component of the water cycle, and is responsible for depositing most of the fresh water on the planet. Approximately 505,000 km^3 (121,000 mi^3) of water falls as precipitation each year, 398,000 km^3 (95,000 cu mi) of it over the oceans. Given the Earth's surface area, that means the globally averaged annual precipitation is 990 millimetres (39 in).

A thunderstorm with heavy precipitation

Mechanisms of producing precipitation include convective, stratiform, and orographic rainfall.Convective processes involve strong vertical motions that can cause the overturning of the atmosphere in that location within an hour and cause heavy precipitation, while stratiform processes involve weaker upward motions and less intense precipitation.

Precipitation can be divided into three categories, based on whether it falls as liquid water, liquid water that freezes on contact with the surface, or ice. Mixtures of different types of precipitation, including types in different categories, can fall simultaneously. Liquid forms of precipitation include rain and drizzle. Rain or drizzle that freezes on contact within a subfreezing air mass is called "freezing rain" or "freezing drizzle". Frozen forms of precipitation include snow, ice needles, ice pellets, hail, and graupel.

Precipitation Types

In meteorology, "precipitation types" can include the character or phase of the precipitation which is falling to ground level. There are three distinct ways that precipitation can occur. Convective precipitation is generally more intense, and of shorter duration, than stratiform precipitation. Orographic precipitation occurs when moist air is forced upwards over rising terrain, such as a mountain.

Precipitation can also fall in either liquid or solid phases, or transition between them. Liquid forms of precipitation include rain and drizzle. Rain or drizzle which freezes on contact within a subfreezing air mass gains the preceding adjective "freezing", becoming known as freezing rain or freezing drizzle. Frozen forms of precipitation include snow, ice needles, sleet, hail, and graupel. Intensity is classified either by rate of fall, or by visibility restriction.

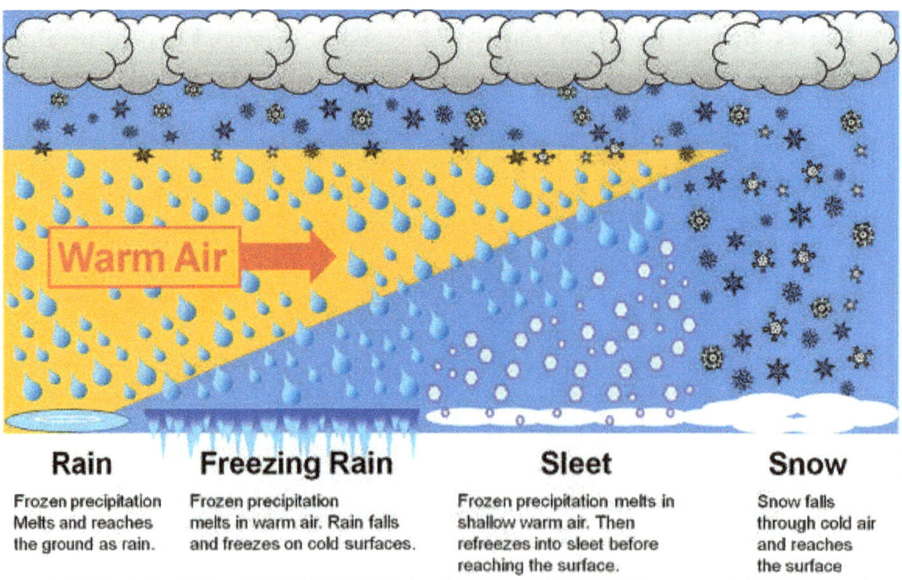

Typical precipitation types associated with a warm front advancing over frigid air

Precipitation in the form of a sunshower

Phases

Precipitation falls in many forms, or phases. They can be subdivided into:

- Liquid precipitation:
  - Drizzle (DZ)
  - Rain (RA)
  - Fog condensation on vegetation foliage, dripping on forest soil.
- Freezing precipitation:
  - Freezing drizzle (FZDZ)
  - Freezing rain (FZRA)
  - Rain and snow mixed / "Snain" (RASN)
  - Sun Shower
- Frozen precipitation:
  - Snow (SN)
  - Snow grains (SG)
  - Ice pellets / Sleet (PL)
  - Hail (GR)
  - Snow pellets / Graupel (GS)
  - Ice crystals (IC)

The parenthesized letters are the METAR codes for each phenomenon.

Mechanisms

Precipitation occurs when the air becomes saturated with water vapor, and can no longer hold all of it in vapor form. This occurs when moist air cools, usually when an airmass rises through the atmosphere. However, an airmass can also cool (e.g. through radiative cooling, or ground contact with cold terrain) without a change in altitude.

Convective precipitation occurs when air rises vertically through the (temporarily) self-sustaining mechanism of convection. *Stratiform* precipitation occurs when large air masses rise diagonally as larger-scale atmospheric dynamics force them to move over each other. *Orographic* precipitation is similar, except the upwards motion is forced when a moving airmass encounters the rising slope of a landform such as a mountain ridge.

Convectional

Convection occurs when the Earth's surface, within a conditionally unstable or moist atmosphere, becomes heated more than its surroundings, leading to significant evaporation. Convective rain, or showery precipitation, occurs from convective clouds, e.g., cumulonimbus or cumulus congestus. It falls as showers with rapidly changing intensity. Convective precipitation falls over a certain area for a relatively short time, as convective clouds have limited horizontal extent. Most precipitation in the tropics appears to be convective; however, it has been suggested that stratiform precipitation thunderstorms.

A Violent electrical storm results from convective cumulus congestus clouds.

Graupel and hail indicate convection. In mid-latitudes, convective precipitation is associated with cold fronts (often behind the front), squall lines, and warm fronts in very moist air.

Cyclonic

Frontal precipitation is caused by frontal systems surrounding extratropical cyclones or lows, which form when warm and often tropical air meets cooler air. Frontal precipitation typically falls out of nimbostratus clouds.

A weather front is the boundary of two airmasses

When masses of air with different density (moisture and temperature characteristics) meet, less dense warmer air overrides colder air. The warmer air is forced to rise and if conditions are right becomes saturated, causing precipitation. In turn, precipitation can enhance the temperature and moisture contrast along a frontal boundary. Passing weather fronts cause sudden changes in general temperature, and in the humidity and pressure in the air at ground level.

Warm fronts occur where the warm air pushes out a previously lodged cold air mass. The warm air overrides the cooler air and moves upward. Warm fronts are followed by extended periods of light rain and drizzle, because, after the warm air rises above the cooler air (which remains on the ground), it gradually cools due to the air's expansion while being lifted, which forms clouds and leads to precipitation.

Cold fronts occur when a mass of cooler air dislodges a mass of warm air. This type of transition is sharper, since cold air is more dense than warm air. The rain duration is shorter, and generally more intense, than that which occurs ahead of warm fronts.

A wide variety of weather can be found along an occluded front, with thunderstorms possible, but usually their passage is associated with a drying of the air mass.

Orographic

Orographic or relief rainfall is caused when masses of air pushed by wind are forced up the side of elevated land formations, such as large mountains.

Orographic precipitation occurs when moist air is forced upwards by terrain.

The lift of the air up the side of the mountain results in adiabatic cooling, and ultimately condensation and precipitation. In mountainous parts of the world subjected to relatively consistent winds (for example, the trade winds), a more moist climate usually prevails on the windward side of a mountain than on the leeward (downwind) side. Moisture is removed by orographic lift, leaving drier air (see Foehn) on the descending (generally warming), leeward side where a rain shadow is observed.

Post orographic lift isolated towering vertical thunderhead shower in Mojave Desert, western USA

In Hawaii, Mount Wai'ale'ale (*Wai'ale'ale*), on the island of Kauai, is notable for its extreme rainfall. It has the highest average annual rainfall on Earth, with 460 inches (12,000 mm) per year. Storm systems affect the region with heavy rains between October and March. Local climates vary considerably on each island due to their topography, divisible into windward (*Ko olau*) and leeward (*Kona*) regions based upon location relative to the higher mountains. Windward sides face the east-to-northeast trade winds and receive much more rainfall; leeward sides are drier and sunnier, with less rain and less cloud cover. On the island of Oahu, clouds and often rain can usually be observed around the windward mountain peaks, while the southern parts of the island (including most of Honolulu and Waikiki) receive dramatically less rainfall throughout the year.

In South America, the Andes mountain range blocks Pacific Ocean moisture that arrives on the continent, resulting in a desert-like climate just downwind across western Argentina. The Sierra Nevada range creates the same effect in North America, causing the Great Basin desert, Mojave Desert, and Sonoran Desert.

Intensity

Precipitation is measured using a rain gauge. When classified according to the rate of precipitation, rain can be divided into categories. Light rain describes rainfall which falls at a rate of between a trace and 2.5 millimetres (0.098 in) per hour. Moderate rain describes rainfall with a precipitation rate of between 2.6 millimetres (0.10 in) and 7.6 millimetres (0.30 in) per hour. Heavy rain describes rainfall with a precipitation rate above 7.6 millimetres (0.30 in) per hour.

Snowfall intensity is classified in terms of visibility. When the visibility is over 1 kilometre (0.62 mi), snow is determined to be light. Moderate snow describes snowfall with visibility restrictions between .5 kilometres (0.31 mi) and 1 kilometre (0.62 mi). Heavy snowfall describes conditions when visibility is restricted below .5 kilometres (0.31 mi).

Gallery

Tropical downpour in Hong Kong

Heavy rain falling in Kolkata, India

Graupel accumulation in Elko, Nevada

HOW THE AIR BECOMES SATURATED

Cooling Air to its Dew Point

The dew point is the temperature to which a parcel must be cooled in order to become saturated, and (unless super-saturation occurs) condenses to water. Water vapour normally begins to condense on condensation nuclei such as dust, ice, and salt in order to form clouds. An elevated portion of a frontal zone forces broad areas of lift, which form clouds decks such as altostratus or cirrostratus. Stratus is a stable cloud deck which tends to form when a cool, stable air mass is trapped underneath a warm air mass. It can also form due to the lifting of advection fog during breezy conditions.

Late-summer rainstorm in Denmark

Lenticular cloud forming due to mountains over Wyoming

There are four main mechanisms for cooling the air to its dew point: adiabatic cooling, conductive cooling, radiational cooling, and evaporative cooling. Adiabatic cooling occurs when air rises and expands. The air can rise due to convection, large-scale atmospheric motions, or a physical barrier such as a mountain (orographic lift). Conductive cooling occurs when the air comes into contact with a colder surface,usually by being blown from one surface to another, for example from a liquid water surface to colder land. Radiational cooling occurs due to the emission of infrared radiation, either by the air or by the surface underneath. Evaporative cooling occurs when moisture is added to the air through evaporation, which forces the air temperature to cool to its wet-bulb temperature, or until it reaches saturation.

Adding Moisture to the Air

The main ways water vapour is added to the air are: wind convergence into areas of upward motion,precipitation or virga falling from above, daytime heating evaporating water from the surface of oceans, water bodies or wet land, transpiration from plants, cool or dry air moving over warmer water, and lifting air over mountains.

FORMATION

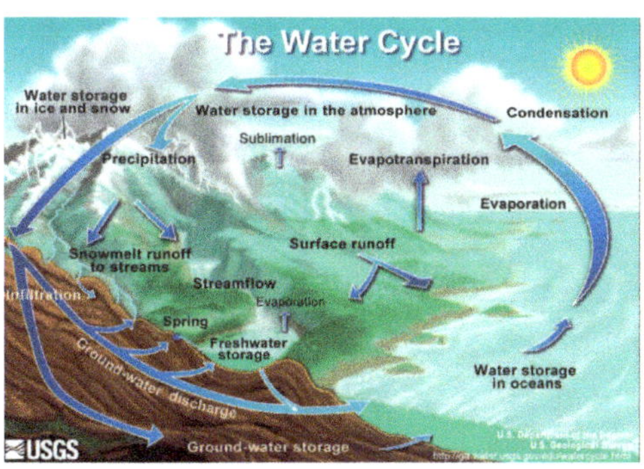

Condensation and coalescence are important parts of the water cycle.

Raindrops

Coalescence occurs when water droplets fuse to create larger water droplets, or when water droplets freeze onto an ice crystal, which is known as the Bergeron process. The fall rate of very small droplets is negligible, hence clouds do not fall out of the sky; precipitation will only occur when these coalesce into larger drops. When air turbulence occurs, water droplets collide, producing larger droplets. As these larger water droplets descend, coalescence continues, so that drops become heavy enough to overcome air resistance and fall as rain.

Raindrops have sizes ranging from 0.1 millimetres (0.0039 in) to 9 millimetres (0.35 in) mean diameter, above which they tend to break up. Smaller drops are called cloud droplets, and their shape is spherical. As a raindrop increases in size, its shape becomes more oblate, with its largest cross-section facing the oncoming airflow. Contrary to the cartoon pictures of raindrops, their shape does not resemble a teardrop. Intensity and duration of rainfall are usually inversely related, i.e., high intensity storms are likely to be of short duration and low intensity storms can have a long duration. Rain drops associated with melting hail tend to be larger than other rain drops. The METAR code for rain is RA, while the coding for rain showers is SHRA.

Ice Pellets

Ice pellets or sleet are a form of precipitation consisting of small, translucent balls of ice. Ice pellets are usually (but not always) smaller than hailstones. They often bounce when they hit the ground, and generally do not freeze into a solid mass unless mixed with freezing rain. The METAR code for ice pellets is **PL**.

An accumulation of ice pellets

Ice pellets form when a layer of above-freezing air exists with sub-freezing air both above and below. This causes the partial or complete melting of any snowflakes falling through the warm layer. As they fall back into the sub-freezing layer closer to the surface, they re-freeze into ice pellets. However, if the sub-freezing layer beneath the warm layer is too small, the precipitation will not have time to re-freeze, and freezing rain will be the result at the surface. A temperature profile showing a warm layer above the ground is most likely to be found in advance of a warm front during the cold season, but can occasionally be found behind a passing cold front.

Hail

Like other precipitation, hail forms in storm clouds when supercooled water droplets freeze on contact with condensation nuclei, such as dust or dirt. The storm's updraft blows the hailstones to the upper part of the cloud. The updraft dissipates and the hailstones fall down, back into the updraft, and are lifted again. Hail has a diameter of 5 millimetres (0.20 in) or more.

A large hailstone, about 6 centimetres (2.4 in) in diameter

Within METAR code, GR is used to indicate larger hail, of a diameter of at least 6.4 millimetres (0.25 in). GR is derived from the French word grêle. Smaller-sized hail, as well as snow pellets, use the coding of GS, which is short for the French word grésil. Stones just larger than golf ball-sized are one of the most frequently reported hail sizes. Hailstones can grow to 15 centimetres (6 in) and weigh more than 500 grams (1 lb). In large hailstones, latent heat released by further freezing may melt the outer shell of the hailstone. The hailstone then may undergo 'wet growth', where the liquid outer shell collects other smaller hailstones. The hailstone gains an ice layer and grows increasingly larger with each ascent. Once a hailstone becomes too heavy to be supported by the storm's updraft, it falls from the cloud.

Snowflakes

Snow crystals form when tiny supercooled cloud droplets (about 10 μm in diameter) freeze. Once a droplet has frozen, it grows in the supersaturated environment. Because water droplets are more numerous than the ice crystals the crystals are able to grow to hundreds of micrometers in size at the expense of the water droplets. This process is known as the Wegener–Bergeron–Findeisen process. The corresponding depletion of water vapour causes the droplets to evaporate, meaning that the ice crystals grow at the droplets' expense. These large crystals are an efficient source of precipitation, since they fall through the atmosphere due to their mass, and may collide and stick together in clusters, or aggregates. These aggregates are snowflakes, and are usually the type of ice particle that falls to the ground. Guinness World Records list the world's largest snowflakes as those of January 1887 at Fort Keogh, Montana; allegedly one measured 38 cm (15 inches) wide. The exact details of the sticking mechanism remain a subject of research.

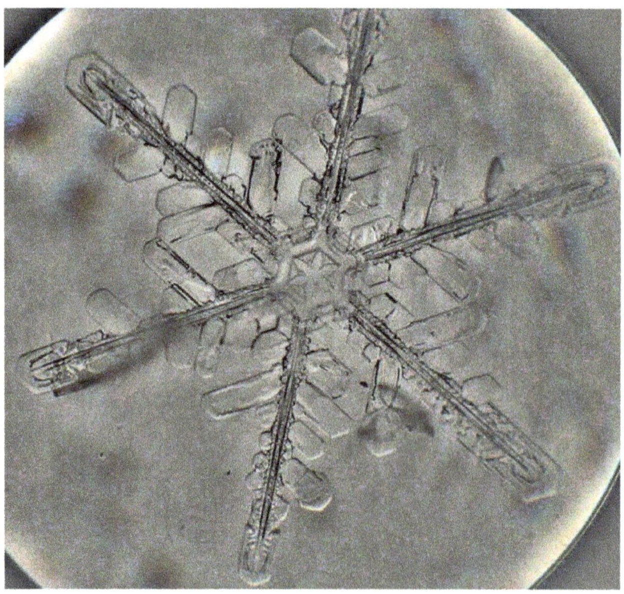

Snowflake viewed in an optical microscope

Although the ice is clear, scattering of light by the crystal facets and hollows/imperfections mean that the crystals often appear white in color due to diffuse reflection of the whole spectrum of light by the small ice particles. The shape of the snowflake is determined broadly by the temperature and humidity at which it is formed. Rarely, at a temperature of around −2 °C (28 °F), snowflakes can form in threefold symmetry—triangular snowflakes. The most common snow particles are visibly irregular, although near-perfect snowflakes may be more common in pictures because they are more visually appealing. No two snowflakes are alike, which grow at different rates and in different patterns depending on the changing temperature and humidity within the atmosphere that the snowflake falls through on its way to the ground.The METAR code for snow is SN, while snow showers are coded SHSN.

Diamond Dust

Diamond dust, also known as ice needles or ice crystals, forms at temperatures approaching −40 °C (−40 °F) due to air with slightly higher moisture from aloft mixing with colder, surface based air. They are made of simple ice crystals that are hexagonal in shape. The METAR identifier for diamond dust within international hourly weather reports is IC.

Causes

Frontal Activity

Stratiform or dynamic precipitation occurs as a consequence of slow ascent of air in synoptic systems (on the order of cm/s), such as over surface cold fronts, and over and ahead of warm fronts. Similar ascent is seen around tropical cyclones outside of the eyewall, and in comma-head precipitation patterns around mid-latitude cyclones. A wide variety of weather can be found along an occluded front, with thunderstorms possible, but usually their passage is associated with a drying of the air mass. Occluded fronts usually form around mature low-pressure areas.Precipitation may occur on celestial bodies other than Earth. When it gets cold, Mars has precipitation that most likely takes the form of ice needles, rather than rain or snow.

Convection

Convective rain, or showery precipitation, occurs from convective clouds, e.g., cumulonimbus or cumulus congestus. It falls as showers with rapidly changing intensity.

Convective precipitation

Convective precipitation falls over a certain area for a relatively short time, as convective clouds have limited horizontal extent. Most precipitation in the tropics appears to be convective; however, it has been suggested that stratiform precipitation also occurs. Graupel and hail indicate convection. In mid-latitudes, convective precipitation is intermittent and often associated with baroclinic boundaries such as cold fronts, squall lines, and warm fronts.

Orographic Effects

Orographic precipitation occurs on the windward side of mountains and is caused by the rising air motion of a large-scale flow of moist air across the mountain ridge, resulting in adiabatic cooling and condensation. In mountainous parts of the world subjected to relatively consistent winds (for example, the trade winds), a more moist climate usually prevails on the windward side of a mountain than on the leeward or downwind side. Moisture is removed by orographic lift, leaving drier air (see katabatic wind) on the descending and generally warming, leeward side where a rain shadow is observed.

Orographic precipitation

In Hawaii, Mount Wai‘ale‘ale, on the island of Kauai, is notable for its extreme rainfall, as it has the second highest average annual rainfall on Earth, with 12,000 millimetres (460 in). Storm systems affect the state with heavy rains between October and March. Local climates vary considerably on each island due to their topography, divisible into windward (*Ko‘olau*) and leeward (*Kona*) regions based upon location relative to the higher mountains. Windward sides face the east to northeast trade winds and receive much more rainfall; leeward sides are drier and sunnier, with less rain and less cloud cover.

In South America, the Andes mountain range blocks Pacific moisture that arrives in that continent, resulting in a desertlike climate just downwind across western Argentina. The Sierra Nevada range creates the same effect in North America forming the Great Basin and Mojave Deserts.

Snow

Extratropical cyclones can bring cold and dangerous conditions with heavy rain and snow with winds exceeding 119 km/h (74 mph), (sometimes referred to as windstorms in Europe). The band of precipitation that is associated with their warm front is often extensive, forced by weak upward vertical motion of air over the frontal boundary which condenses as it cools and produces precipitation within an elongated band, which is wide and stratiform, meaning falling out of nimbostratus clouds. When moist air tries to dislodge an arctic air mass, overrunning snow can result within the poleward side of the elongated precipitation band. In the Northern Hemisphere, poleward is towards the North Pole, or north. Within the Southern Hemisphere, poleward is towards the South Pole, or south.

Lake-effect snow bands near the Korean Peninsula

Southwest of extratropical cyclones, curved cyclonic flow bringing cold air across the relatively warm water bodies can lead to narrow lake-effect snow bands. Those bands bring strong localized snowfall which can be understood as follows: Large water bodies such as lakes efficiently store heat that results in significant temperature differences (larger than 13 °C or 23 °F) between the water surface and the air above. Because of this temperature difference, warmth and moisture are transported upward, condensing into vertically oriented clouds (see satellite picture) which produce snow showers. The temperature decrease with height and cloud depth are directly affected by both the water temperature and the large-scale environment. The stronger the temperature decrease with height, the deeper the clouds get, and the greater the precipitation rate becomes.

In mountainous areas, heavy snowfall accumulates when air is forced to ascend the mountains and squeeze out precipitation along their windward slopes, which in cold conditions, falls in the form of snow. Because of the ruggedness of terrain, forecasting the location of heavy snowfall remains a significant challenge.

Within the Tropics

The wet, or rainy, season is the time of year, covering one or more months, when most of the average annual rainfall in a region falls. The term *green season* is also sometimes used as a euphemism by tourist authorities. Areas with wet seasons are dispersed across portions of the tropics and subtropics. Savanna climates and areas with monsoon regimes have wet summers and dry winters. Tropical rainforests technically do not have dry or wet seasons, since their rainfall is equally distributed through the year. Some areas with pronounced rainy seasons will see a break in rainfall mid-season when the intertropical convergence zone or monsoon trough move poleward of their location during the middle of the warm season. When the wet season occurs during the warm season, or summer, rain falls mainly during the late afternoon and early evening hours. The wet season is a time when air quality improves, freshwater quality improves, and vegetation grows significantly. Soil nutrients diminish and erosion increases. Animals have adaptation and survival strategies for the wetter regime. The previous dry season leads to food shortages into the wet season, as the crops have yet to mature. Developing countries have noted that their populations show seasonal weight fluctuations due to food shortages seen before the first harvest, which occurs late in the wet season.

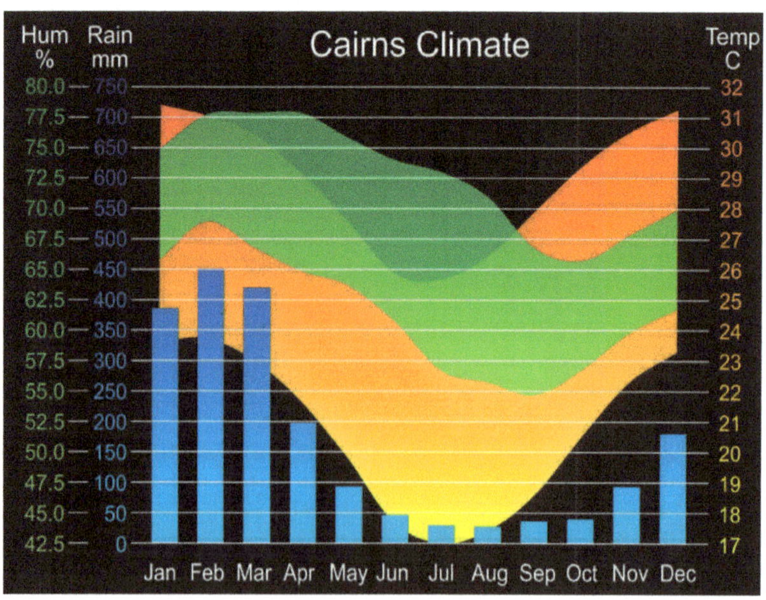

Rainfall distribution by month in Cairns showing the extent of the wet season at that location

Tropical cyclones, a source of very heavy rainfall, consist of large air masses several hundred miles across with low pressure at the centre and with winds blowing inward towards the centre in either a clockwise direction (southern hemisphere) or counterclockwise (northern hemisphere).Although cyclones can take an enormous toll in lives and personal property, they may be important factors in the precipitation regimes of places they impact, as they may bring much-needed precipitation to otherwise dry regions. Areas in their path can receive a year's worth of rainfall from a tropical cyclone passage.

Large-Scale Geographical Distribution

On the large scale, the highest precipitation amounts outside topography fall in the tropics, closely tied to the Intertropical Convergence Zone, itself the ascending branch of the Hadley cell. Mountainous locales near the equator in Colombia are amongst the wettest places on Earth. North and south of this are regions of descending air that form subtropical ridges where precipitation is low; the land surface underneath is usually arid, which forms most of the Earth's deserts. An exception to this rule is in Hawaii, where upslope flow due to the trade winds lead to one of the wettest locations on Earth. Otherwise, the flow of the Westerlies into the Rocky Mountains lead to the wettest, and at elevation snowiest,locations within North America. In Asia during the wet season, the flow of moist air into the Himalayas leads to some of the greatest rainfall amounts measured on Earth in northeast India.

WEATHER FRONT

A weather (state of atmosphere) front is a boundary separating two masses of air of different densities, and is the principal cause of meteorological phenomena outside the tropics. In surface weather analyses, fronts are depicted using various colored triangles and half-circles, depending on the type of front. The air masses separated by a front usually differ in temperature and humidity. Cold fronts may feature narrow bands of thunderstorms and severe weather, and may on occasion be preceded by squall lines or dry lines. Warm fronts are usually preceded by stratiform precipitation and fog. The weather usually clears quickly after a front's passage. Some fronts produce no precipitation and little cloudiness, although there is invariably a wind shift.

Cold fronts and occluded fronts generally move from west to east, while warm fronts move poleward. Because of the greater density of air in their wake, cold fronts and cold occlusions move faster than warm fronts and warm occlusions. Mountains and warm bodies of water can slow the movement of fronts. When a front becomes stationary, and the density contrast across the frontal boundary vanishes, the front can degenerate into a line which separates regions of differing wind velocity, known as a shearline. This is most common over the open ocean.

Approaching weather fronts are often visible from the ground, but are not always as well defined as this

Bergeron Classification of Air Masses

The Bergeron classification is the most widely accepted form of air mass classification. Air mass classification involves three letters. The first letter describes its moisture properties, with c used for continental air masses (dry) and m for maritime air masses (moist). The second letter describes the thermal characteristic of its source region: T for tropical, P for polar, A for arctic or Antarctic, M for monsoon, E for equatorial, and S for superior air (dry air formed by significant upward motion in the atmosphere). The third letter is used to designate the stability of the atmosphere. If the air mass is colder than the ground below it, it is labeled k. If the air mass is warmer than the ground below it, it is labeled w. Fronts separate air masses of different types or origins, and are located along troughs of lower pressure.

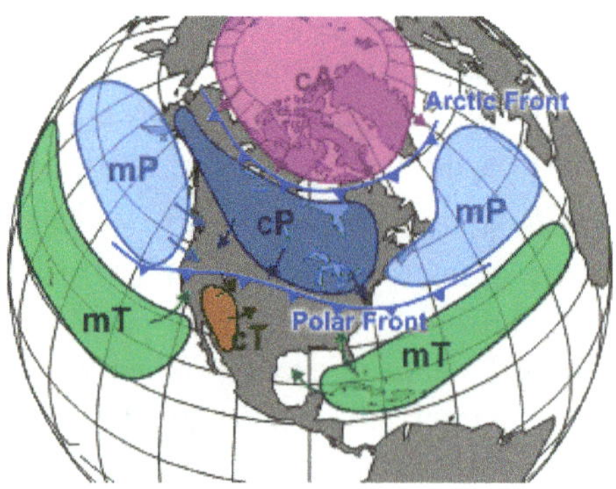

Different air masses which affect North America, as well as other continents, tend to be separated by frontal boundaries. In this illustration, the Arctic front separates Arctic from Polar air masses, while the Polar front separates Polar air from warm air masses. (cA is continental arctic; cP is continental polar; mP is maritime polar; cT is continental tropic; and mT is maritime tropic.)

Surface Weather Analysis

A surface weather analysis is a special type of weather map which provides a view of weather elements over a geographical area at a specified time based on information from ground-based weather stations. Weather maps are created by plotting or tracing the values of relevant quantities such as sea-level pressure, temperature, and cloud cover onto a geographical map to help find synoptic scale features such as weather fronts.

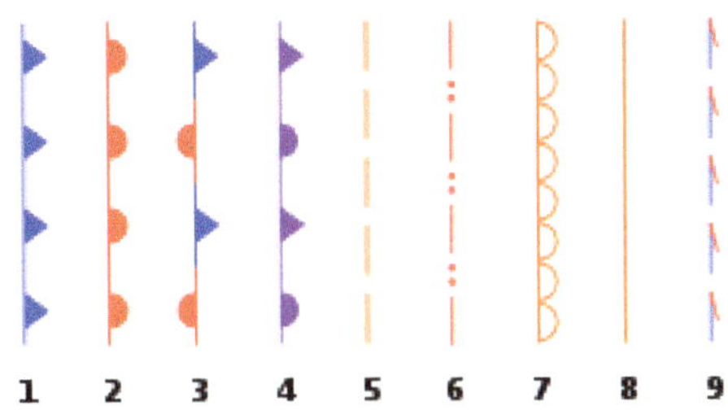

Weather map symbols:

1. cold front;
2. warm front;
3. stationary front;
4. occluded front;
5. surface trough;
6. squall/shear line;
7. dry line;
8. tropical wave;
9. trowal

Surface weather analyses have special symbols which show frontal systems, cloud cover, precipitation, or other important information. For example, an *H* may represent high pressure, implying fair weather. An *L* on the other hand may represent low pressure, which frequently accompanies precipitation. Low pressure also creates surface winds deriving from high pressure zones. Various symbols are used not just for frontal zones and other surface boundaries on weather maps, but also to depict the present weather at various locations on the weather map. In addition, areas of precipitation help determine the frontal type and location.

Front Types

There are two different meanings used within meteorology to describe weather around a frontal zone. The term "anafront" describes boundaries which show instability, meaning air rises rapidly along and over the boundary to cause significant weather changes. A "katafront" is weaker, bringing smaller changes in temperature and moisture, as well as limited rainfall.

Cold Front

A cold front is located at the leading edge of the temperature drop off, which in an isotherm analysis shows up as the leading edge of the isotherm gradient, and it normally lies within a sharp surface trough. Cold fronts often bring heavy thunderstorms, rain and hail. Cold front can produce sharper changes in weather and move up to twice as quickly as warm fronts, since cold air is denser than warm air and rapidly replaces the warm air preceding the boundary. On weather maps, the surface position of the cold front is marked with the symbol of a blue line of triangle-shaped pips pointing in the direction of travel, and it is placed at the leading edge of the cooler air mass. Cold fronts come in association with a low-pressure area. The concept of colder, dense air "wedging" under the less dense warmer air is often used to depict how air is lifted along a frontal boundary. The cold air wedging underneath warmer air creates the strongest winds just above the ground surface, a phenomenon often associated with property-damaging wind gusts. This lift would then form a narrow line of showers and thunderstorms if enough moisture were present. However, this concept isn't an accurate description of the physical processes; upward motion is not produced because of warm air "ramping up" cold, dense air, rather, frontogenetical circulation is behind the upward forcing.

Warm Front

Warm fronts are at the leading edge of a homogeneous warm air mass, which is located on the equatorward edge of the gradient in isotherms, and lie within broader troughs of low pressure than cold fronts. A warm front moves more slowly than the cold front which usually follows because cold air is denser and harder to remove from the Earth's surface. This also forces temperature differences across warm fronts to be broader in scale. Clouds ahead of the warm front are mostly stratiform, and rainfall gradually increases as the front approaches. Fog can also occur preceding a warm frontal passage. Clearing and warming is usually rapid after frontal passage. If the warm air mass is unstable, thunderstorms may be embedded among the stratiform clouds ahead of the front, and after frontal passage thundershowers may continue. On weather maps, the surface location of a warm front is marked with a red line of semicircles pointing in the direction of travel.

Occluded Front

An occluded front is formed when a cold front overtakes a warm front. and usually form around mature low-pressure areas. The cold and warm fronts curve naturally poleward into the point of occlusion, which is also known as the triple point. It lies within a sharp trough, but the air mass behind the boundary can be either warm or cold. In a cold occlusion, the air mass overtaking the warm front is cooler than the cool air ahead of the warm front and plows under both air masses. In a warm occlusion, the air mass overtaking the warm front is warmer than the cold air ahead of the warm front and rides over the colder air mass while lifting the warm air.

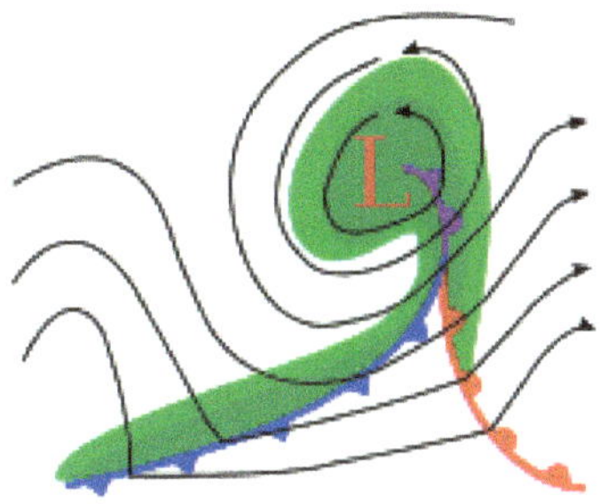

Occluded front depiction for the Northern Hemisphere

A wide variety of weather can be found along an occluded front, with thunderstorms possible, but usually their passage is associated with a drying of the air mass. Within the occlusion of the front, a circulation of air brings warm air upward and sends drafts of cold air downward, or vice versa depending on the occlusion the front is experiencing. Precipitations and clouds are associated with the *trowal*, the projection on the Earth's surface of the tongue of warm air aloft formed during the occlusion process of the depression.

Occluded fronts are indicated on a weather map by a purple line with alternating half-circles and triangles pointing in direction of travel. The trowal is indicated by a series of blue and red junction lines.

Stationary Front

A stationary front is a non-moving (or stalled) boundary between two air masses, neither of which is strong enough to replace the other. They tend to remain essentially in the same area for extended periods of time, usually moving in waves. There is normally a broad temperature gradient behind the boundary with more widely spaced isotherm packing.

A wide variety of weather can be found along a stationary front, but usually clouds and prolonged precipitation are found there. Stationary fronts either dissipate after several days or devolve into shear lines, but they can transform into a cold or warm front if conditions aloft change. Stationary fronts are marked on weather maps with alternating red half-circles and blue spikes pointing in opposite directions, indicating no significant movement.

When stationary fronts become smaller in scale, degenerating to a narrow zone where wind direction changes significantly over a relatively short distance, they become known as shearlines. A shearline is depicted as a line of red dots and dashes.

Dry Line

A similar phenomenon to a weather front is the dry line, which is the boundary between air masses with significant moisture differences. When the westerlies increase on the north side of surface highs, areas of lowered pressure will form downwind of north–south oriented mountain chains, leading to the formation of a lee trough. Near the surface during daylight hours, warm moist air is denser than dry air of greater temperature, and thus the warm moist air wedges under the drier air like a cold front. At higher altitudes, the warm moist air is less dense than the dry air and the boundary slope reverses. In the vicinity of the reversal aloft, severe weather is possible, especially when a triple point is formed with a cold front.A weaker form of the dry line seen more commonly is the lee trough, which displays weaker differences in moisture. When moisture pools along the boundary during the warm season, it can be the focus of diurnal thunderstorms.

The dry line may occur anywhere on earth in regions intermediate between desert areas and warm seas. The southern plains west of the Mississippi River in the United States are a particularly favored location. The dry line normally moves eastward during the day and westward at night. A dry line is depicted on National Weather Service (NWS) surface analyses as an orange line with scallops facing into the moist sector. Dry lines are one of the few surface fronts where the pips indicated do not necessarily reflect the direction of motion.

Squall Line

Organized areas of thunderstorm activity not only reinforce pre-existing frontal zones, but can outrun cold fronts in a pattern where the upper level jet splits apart into two streams, with the resultant Mesoscale Convective System (MCS) forming at the point of the upper level split in the wind pattern running southeast into the warm sector parallel to low-level thickness lines. When the convection is strong and linear or curved, the MCS is called a squall line, with the feature placed at the leading edge of the significant wind shift and pressure rise. Even weaker and less organized areas of thunderstorms lead to locally cooler air and higher pressures, and outflow boundaries exist ahead of this type of activity, which can act as foci for additional thunderstorm activity later in the day.

A shelf cloud such as this one can be a sign that a squall is imminent

These features are often depicted in the warm season across the United States on surface analyses and lie within surface troughs. If outflow boundaries or squall lines form over arid regions, a haboob may result. Squall lines are depicted on NWS surface analyses as an alternating pattern of two red dots and a dash labelled SQLN or SQUALL LINE, while outflow boundaries are depicted as troughs with a label of OUTFLOW BOUNDARY.

Precipitation Produced

Fronts are the principal cause of significant weather. *Convective precipitation* (showers, thundershowers, and related unstable weather) is caused by air being lifted and condensing into clouds by the movement of the cold front or cold occlusion under a mass of warmer, moist air. If the temperature differences of the two air masses involved are large and the turbulence is extreme because of wind shear and the presence of a strong jet stream, "roll clouds" and tornadoes may occur.

Convective precipitation

In the warm season, lee troughs, breezes, outflow boundaries and occlusions can lead to convection if enough moisture is available. *Orographic precipitation* is precipitation created through the lifting action of air moving over terrain such as mountains and hills, which is most common behind cold fronts that move into mountainous areas. It may sometimes occur in advance of warm fronts moving northward to the east of mountainous terrain. However, precipitation along warm fronts is relatively steady, as in rain or drizzle. Fog, sometimes extensive and dense, often occurs in pre-warm-frontal areas. Although, not all fronts produce precipitation or even clouds because moisture must be present in the air mass which is being lifted.

Movement

Fronts are generally guided by winds aloft, but do not move as quickly. Cold fronts and occluded fronts in the Northern Hemisphere usually travel from the northwest to southeast, while warm fronts move more poleward with time. In the Northern Hemisphere a warm front moves from southwest to northeast. In the Southern Hemisphere, the reverse is true; a cold front usually moves from southwest to northeast, and a warm front moves from northwest to southeast. Movement is largely caused by the pressure gradient force (horizontal differences in atmospheric pressure) and the Coriolis effect, which is caused by Earth's spinning about its axis. Frontal zones can be slowed down by geographic features like mountains and large bodies of warm water.

Measurement

The standard way of measuring rainfall or snowfall is the standard rain gauge, which can be found in 100 mm (4 in) plastic and 200 mm (8 in) metal varieties. The inner cylinder is filled by 25 mm (1 in) of rain, with overflow flowing into the outer cylinder. Plastic gauges have markings on the inner cylinder down to 0.25 mm (0.01 in) resolution, while metal gauges require use of a stick designed with the appropriate 0.25 mm (0.01 in) markings. After the inner cylinder is filled, the amount inside it is discarded, then filled with the remaining rainfall in the outer cylinder until all the fluid in the outer cylinder is gone, adding to the overall total until the outer cylinder is empty. These gauges are used in the winter by removing the funnel and inner cylinder and allowing snow and freezing rain to collect inside the outer cylinder. Some add anti-freeze to their gauge so they do not have to melt the snow or ice that falls into the gauge. Once the snowfall/ice is finished accumulating, or as 300 mm (12 in) is approached, one can either bring it inside to melt, or use lukewarm water to fill the inner cylinder with in order to melt the frozen precipitation in the outer cylinder, keeping track of the warm fluid added, which is subsequently subtracted from the overall total once all the ice/snow is melted.

Standard rain gauge

Other types of gauges include the popular wedge gauge (the cheapest rain gauge and most fragile), the tipping bucket rain gauge, and the weighing rain gauge. The wedge and tipping bucket gauges will have problems with snow. Attempts to compensate for snow/ice by warming the tipping bucket meet with limited success, since snow may sublimate if the gauge is kept much above freezing. Weighing gauges with antifreeze should do fine with snow, but again, the funnel needs to be removed before the event begins. For those looking to measure rainfall the most inexpensively, a can that is cylindrical with straight sides will act as a rain gauge if left out in the open, but its accuracy will depend on what ruler is used to measure the rain with. Any of the above rain gauges can be made at home, with enough know-how.

When a precipitation measurement is made, various networks exist across the United States and elsewhere where rainfall measurements can be submitted through the Internet, such as CoCoRAHS or GLOBE. If a network is not available in the area where one lives, the nearest local weather office will likely be interested in the measurement.

Hydrometeor Definition

A concept used in precipitation measurement is the hydrometeor. Bits of liquid or solid water in the atmosphere are known as hydrometeors. Formations due to condensation, such as clouds, haze, fog, and mist, are composed of hydrometeors. All precipitation types are made up of hydrometeors by definition, including virga, which is precipitation which evaporates before reaching the ground. Particles blown from the Earth's surface by wind, such as blowing snow and blowing sea spray, are also hydrometeors.

Satellite Estimates

Although surface precipitation gauges are considered the standard for measuring precipitation, there are many areas in which their use is not feasible. This includes the vast expanses of ocean and remote land areas. In other cases, social, technical or administrative issues prevent the dissemination of gauge observations. As a result, the modern global record of precipitation largely depends on satellite observations.

Satellite sensors work by remotely sensing precipitation—recording various parts of the electromagnetic spectrum that theory and practice show are related to the occurrence and intensity of precipitation. The sensors are almost exclusively passive, recording what they see, similar to a camera, in contrast to active sensors (radar, lidar) that send out a signal and detect its impact on the area being observed.

Satellite sensors now in practical use for precipitation fall into two categories. Thermal infrared (IR) sensors record a channel around 11 micron wavelength and primarily give information about cloud tops. Due to the typical structure of the atmosphere, cloud-top temperatures are approximately inversely related to cloud-top heights, meaning colder clouds almost always occur at higher altitudes. Further, cloud tops with a lot of small-scale variation are likely to be more vigorous than smooth-topped clouds. Various mathematical schemes, or algorithms, use these and other properties to estimate precipitation from the IR data.

The second category of sensor channels is in the microwave part of the electromagnetic spectrum. The frequencies in use range from about 10 gigahertz to a few hundred GHz. Channels up to about 37 GHz primarily provide information on the liquid hydrometeors (rain and drizzle) in the lower parts of clouds, with larger amounts of liquid emitting higher amounts of microwave radiant energy. Channels above 37 GHz display emission signals, but are dominated by the action of solid hydrometeors (snow, graupel, etc.) to scatter microwave radiant energy. Satellites such as the Tropical Rainfall Measuring Mission (TRMM) and the Global Precipitation Measurement (GPM) mission employ microwave sensors to form precipitation estimates.

Additional sensor channels and products have been demonstrated to provide additional useful information including visible channels, additional IR channels, water vapor channels and atmospheric sounding retrievals. However, most precipitation data sets in current use do not employ these data sources.

Satellite Data Sets

The IR estimates have rather low skill at short time and space scales, but are available very frequently (15 minutes or more often) from satellites in geosynchronous Earth orbit. IR works best in cases of deep, vigorous convection—such as the tropics—and becomes progressively less useful in areas where stratiform (layered) precipitation dominates, especially in mid- and high-latitude regions. The more-direct physical connection between hydrometeors and microwave channels gives the microwave estimates greater skill on short time and space scales than is true for IR. However, microwave sensors fly only on low Earth orbit satellites, and there are few enough of them that the average time between observations exceeds three hours. This several-hour interval is insufficient to adequately document precipitation because of the transient nature of most precipitation systems as well as the inability of a single satellite to appropriately capture the typical daily cycle of precipitation at a given location.

Since the late 1990s, several algorithms have been developed to combine precipitation data from multiple satellites' sensors, seeking to emphasize the strengths and minimize the weaknesses of the individual input data sets. The goal is to provide "best" estimates of precipitation on a uniform time/space grid, usually for as much of the globe as possible. In some cases the long-term homogeneity of the dataset is emphasized, which is the Climate Data Record standard.

In other cases, the goal is producing the best instantaneous satellite estimate, which is the High Resolution Precipitation Product approach. In either case, of course, the less-emphasized goal is also considered desirable. One key result of the multi-satellite studies is that including even a small amount of surface gauge data is very useful for controlling the biases that are endemic to satellite estimates. The difficulties in using gauge data are that 1) their availability is limited, as noted above, and 2) the best analyses of gauge data take two months or more after the observation time to undergo the necessary transmission, assembly, processing and quality control. Thus, precipitation estimates that include gauge data tend to be produced further after the observation time than the no-gauge estimates. As a result, while estimates that include gauge data may provide a more accurate depiction of the "true" precipitation, they are generally not suited for real- or near-real-time applications.

The work described has resulted in a variety of datasets possessing different formats, time/space grids, periods of record and regions of coverage, input datasets, and analysis procedures, as well as many different forms of dataset version designators. In many cases, one of the modern multi-satellite data sets is the best choice for general use.

Return Period

The likelihood or probability of an event with a specified intensity and duration, is called the return period or frequency. The intensity of a storm can be predicted for any return period and storm duration, from charts based on historic data for the location. The term *1 in 10 year storm* describes a rainfall event which is rare and is only likely to occur once every 10 years, so it has a 10 percent likelihood any given year. The rainfall will be greater and the flooding will be worse than the worst storm expected in any single year. The term *1 in 100 year storm* describes a rainfall event which is extremely rare and which will occur with a likelihood of only once in a century, so has a 1 percent likelihood in any given year. The rainfall will be extreme and flooding to be worse than a 1 in 10 year event. As with all probability events, it is possible though unlikely to have two "1 in 100 Year Storms" in a single year.

Role in Köppen Climate Classification

The Köppen classification depends on average monthly values of temperature and precipitation. The most commonly used form of the Köppen classification has five primary types labeled A through E. Specifically, the primary types are A, tropical; B, dry; C, mild mid-latitude; D, cold mid-latitude; and E, polar. The five primary classifications can be further divided into secondary classifications such as rain forest, monsoon, tropical savanna, humid subtropical, humid continental, oceanic climate, Mediterranean climate, steppe, subarctic climate, tundra, polar ice cap, and desert.

Rain forests are characterized by high rainfall, with definitions setting minimum normal annual rainfall between 1,750 and 2,000 mm (69 and 79 in). A tropical savanna is a grassland biome located in semi-arid to semi-humid climate regions of subtropical and tropical latitudes, with rainfall between 750 and 1,270 mm (30 and 50 in) a year. They are widespread on Africa, and are also found in India, the northern parts of South America, Malaysia, and Australia. The humid subtropical climate zone is where winter rainfall (and sometimes snowfall) is associated with large storms that the westerlies steer from west to east. Most summer rainfall occurs during thunderstorms and from occasional tropical cyclones. Humid subtropical climates lie on the east side continents, roughly between latitudes 20° and 40° degrees away from the equator.

An oceanic (or maritime) climate is typically found along the west coasts at the middle latitudes of all the world's continents, bordering cool oceans, as well as southeastern Australia, and is accompanied by plentiful precipitation year-round.The Mediterranean climate regime resembles the climate of the lands in the Mediterranean Basin, parts of western North America, parts of Western and South Australia, in southwestern South

Africa and in parts of central Chile. The climate is characterized by hot, dry summers and cool, wet winters. A steppe is a dry grassland. Subarctic climates are cold with continuous permafrost and little precipitation.

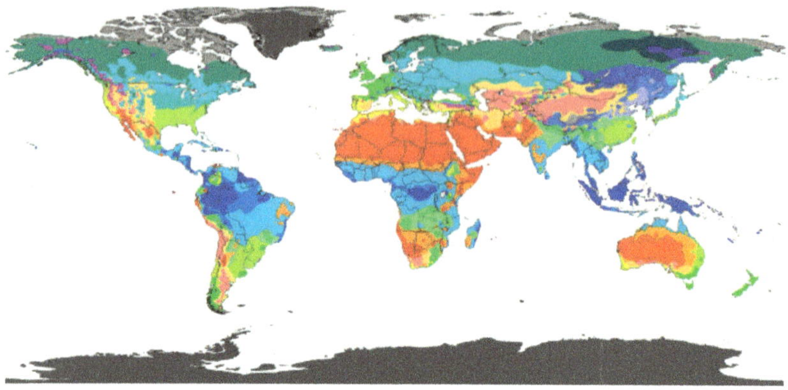

Updated Köppen-Geiger climate map[

KÖPPEN CLIMATE CLASSIFICATION

Köppen climate classification is one of the most widely used climate classification systems. It was first published by Russian German climatologist Wladimir Köppen in 1884, with several later modifications by Köppen, notably in 1918 and 1936. Later, German climatologist Rudolf Geiger (1954, 1961) collaborated with Köppen on changes to the classification system, which is thus sometimes called the Köppen–Geiger climate classification system.

The Köppen climate classification system has been further modified, within the Trewartha climate classification system in the middle 1960s (revised in 1980). The Trewartha system sought to create a more refined middle latitude climate zone, which was one of the criticisms of the Köppen system (the C climate group was too broad).:200–1

Meaning of Symbols

The Köppen climate classification scheme divides climates into five main climate groups: A (tropical), B (dry), C (temperate), D (continental), and E (polar). The second letter indicates the seasonal precipitation type, while the third letter indicates the level of summer heat.

Group A: Tropical (megathermal) climates

This type of climate has every month of the year with an average temperature of 18 °C (64.4 °F) or higher, with significant precipitation.s

- *Af* = Tropical rainforest climate; average precipitation of at least 60 mm (2.4 in) in every month.

- *Am* = Tropical monsoon climate; driest month (which nearly always occurs at or soon after the "winter" solstice for that side of the equator) with preecipitation less than 60 mm (2.4 in), but more than 4% the total annual precipitation.
- *Aw* or *As* = Tropical wet and dry or savanna climate; with the driest month having precipitation less than 60 mm (2.4 in) and less than 4% of the total annual precipitation.

Group B: Dry (arid and semiarid) climates

This type of climate is defined by little precipitation.

Multiply the average annual temperature in Celsius by 20, then add

(a) 280 if 70% or more of the total precipitation is in the spring and summer months (April–September in the Northern Hemisphere, or October–March in the Southern), or

(b) 140 if 30%–70% of the total precipitation is received during the spring and summer, or

(c) 0 if less than 30% of the total precipitation is so received.

If the annual precipitation is less than 50% of this threshold, the classification is BW (arid: desert climate); if it is in the range of 50%–100% of the threshold, the classification is BS (semi-arid: steppe climate).

A third letter can be included to indicate temperature. Originally, h signified low-latitude climate (average annual temperature above 18 °C (64.4 °F)) while k signified middle-latitude climate (average annual temperature below 18 °C), but the more common practice today, especially in the United States, is to use h to mean the coldest month has an average temperature above 0 °C (32 °F), with k denoting that at least one month's averages below 0 °C.

- *BWh* = Hot desert climate
- *BWk* = Cold desert climate
- *BSh* = Hot semi-arid climate
- *BSk* = Cold semi-arid climate

Group C: Temperate (mesothermal) climates

This type of climate has the coldest month averaging between 0 °C (32 °F) and 18 °C (64.4 °F) and at least one month averaging above 10 °C (50 °F).

- *Cfa* = Humid subtropical climate; coldest month averaging above 0 °C (32 °F) and at least one month's average temperature above 22 °C (71.6 °F). No significant precipitation difference between seasons (neither abovementioned set of conditions fulfilled).
- *Cfb* = Temperate oceanic climate; coldest month averaging above 0 °C (32 °F), all months with average temperatures below 22 °C (71.6 °F), and at least four months averaging above 10 °C (50 °F). No significant precipitation difference between seasons (neither abovementioned set of conditions fulfilled).

- *Cfc* = Subpolar oceanic climate; coldest month averaging above 0 °C (32 °F) and 1–3 months averaging above 10 °C (50 °F). No significant precipitation difference between seasons (neither abovementioned set of conditions fulfilled).
- *Cwa* = Monsoon-influenced humid subtropical climate; coldest month averaging above 0 °C (32 °F) and at least one month's average temperature above 22 °C (71.6 °F). At least ten times as much rain in the wettest month of summer as the in the driest month of winter (alternative definition is 70% or more of average annual precipitation is received in the warmest six months).
- *Cwb* = Subtropical highland climate or temperate oceanic climate with dry winters; coldest month averaging above 0 °C (32 °F), all months with average temperatures below 22 °C (71.6 °F), and at least four months averaging above 10 °C (50 °F). At least ten times as much rain in the wettest month of summer as the in the driest month of winter (an alternative definition is 70% or more of average annual precipitation received in the warmest six months).
- *Cwc* = Cold subtropical highland climate or subpolar oceanic climate with dry winters; coldest month averaging above 0 °C (32 °F) and 1–3 months averaging above 10 °C (50 °F). At least ten times as much rain in the wettest month of summer as the in the driest month of winter (alternative definition is 70% or more of average annual precipitation is received in the warmest six months).
- *Csa* = Hot-summer Mediterranean climate; coldest month averaging above 0 °C (32 °F) and at least one month's average temperature above 22 °C (71.6 °F). At least three times as much precipitation in the wettest month of winter as in the driest month of summer, and driest month of summer receives less than 30 mm (1.2 in).
- *Csb* = Warm-summer Mediterranean climate; coldest month averaging above 0 °C (32 °F), all months with average temperatures below 22 °C (71.6 °F), and at least four months averaging above 10 °C (50 °F). At least three times as much precipitation in the wettest month of winter as in the driest month of summer, and driest month of summer receives less than 30 mm (1.2 in).
- *Csc* = Cool-summer Mediterranean climate; coldest month averaging above 0 °C (32 °F) and 1–3 months averaging above 10 °C (50 °F). At least three times as much precipitation in the wettest month of winter as in the driest month of summer, and driest month of summer receives less than 30 mm (1.2 in).

Group D: Continental (microthermal) climates

This type of climate has at least one month averaging below 0 °C (32 °F) and at least one month averaging above 10 °C (50 °F).

- *Dfa* = Hot-summer humid continental climate; coldest month averaging below 0 °C (32 °F) and at least one month's average temperature above 22 °C (71.6 °F). No significant precipitation difference between seasons (neither abovementioned set of conditions fulfilled).
- *Dfb* = Warm-summer humid continental climate; coldest month averaging below 0 °C (32 °F), all months with average temperatures below 22 °C (71.6 °F), and

at least four months averaging above 10 °C (50 °F). No significant precipitation difference between seasons (neither abovementioned set of conditions fulfilled).

- *Dfc* = Subarctic climate; coldest month averaging below 0 °C (32 °F) and 1–3 months averaging above 10 °C (50 °F). No significant precipitation difference between seasons (neither abovementioned set of conditions fulfilled).
- *Dfd* = Extremely cold subarctic climate; coldest month averaging below −38 °C (−36.4 °F) and 1–3 months averaging above 10 °C (50 °F). No significant precipitation difference between seasons (neither abovementioned set of conditions fulfilled).
- *Dwa* = Monsoon-influenced hot-summer humid continental climate; coldest month averaging below 0 °C (32 °F) and at least one month's average temperature above 22 °C (71.6 °F). At least ten times as much rain in the wettest month of summer as the in the driest month of winter (alternative definition is 70% or more of average annual precipitation is received in the warmest six months).
- *Dwb* = Monsoon-influenced warm-summer humid continental climate; coldest month averaging below 0 °C (32 °F), all months with average temperatures below 22 °C (71.6 °F), and at least four months averaging above 10 °C (50 °F). At least ten times as much rain in the wettest month of summer as the in the driest month of winter (alternative definition is 70% or more of average annual precipitation is received in the warmest six months).
- *Dwc* = Monsoon-influenced subarctic climate; coldest month averaging below 0 °C (32 °F) and 1–3 months averaging above 10 °C (50 °F). At least ten times as much rain in the wettest month of summer as the in the driest month of winter (alternative definition is 70% or more of average annual precipitation is received in the warmest six months).
- *Dwd* = Monsoon-influenced extremely cold subarctic climate; coldest month averaging below −38 °C (−36.4 °F) and 1–3 months averaging above 10 °C (50 °F). At least ten times as much rain in the wettest month of summer as the in the driest month of winter (alternative definition is 70% or more of average annual precipitation is received in the warmest six months).
- *Dsa* = Hot, dry-summer continental climate; coldest month averaging below 0 °C (32 °F) and at least one month's average temperature above 22 °C (71.6 °F). At least three times as much precipitation in the wettest month of winter as in the driest month of summer, and driest month of summer receives less than 30 mm (1.2 in).
- *Dsb* = Warm, dry-summer continental climate; coldest month averaging below 0 °C (32 °F), all months with average temperatures below 22 °C (71.6 °F), and at least four months averaging above 10 °C (50 °F). At least three times as much precipitation in the wettest month of winter as in the driest month of summer, and driest month of summer receives less than 30 mm (1.2 in).
- *Dsc* = dry-summer subarctic climate; coldest month averaging below 0 °C (32 °F) and 1–3 months averaging above 10 °C (50 °F). At least three times as much precipitation in the wettest month of winter as in the driest month of summer, and driest month of summer receives less than 30 mm (1.2 in).

- *Dsd* = Extremely cold, dry-summer subarctic climate; coldest month averaging below −38 °C (−36.4 °F) and 1–3 months averaging above 10 °C (50 °F). At least three times as much precipitation in the wettest month of winter as in the driest month of summer, and driest month of summer receives less than 30 mm (1.2 in).

Group E: Polar and alpine (montane) climates

This type of climate has every month of the year with an average temperature below 10 °C (50 °F).

- *ET* = Mild tundra climate; all 12 months of the year with average temperatures between 0 °C (32 °F) and 10 °C (50 °F).
- *ETf* = Cold tundra climate; at least one month with an average temperature below 0 °C (32 °F).
- *EF* = Ice cap climate; all 12 months of the year with average temperatures below 0 °C (32 °F).

Scheme

Group A: Tropical/Megathermal Climates

Tropical climates are characterized by constant high temperatures (at sea level and low elevations); all 12 months of the year have average temperatures of 18 °C (64.4 °F) or higher. They are subdivided as follows:

Tropical Rainforest Climate

All 12 months have average precipitation of at least 60 mm (2.4 in). These climates usually occur within 10° latitude of the equator. This climate is dominated by the doldrums low-pressure system all year round, so has no natural seasons in terms of thermal and moisture changes.

- Examples:
  - Apia, Samoa (*Af*)
  - Bogor, Indonesia (*Af*)
  - Davao, Philippines (*Af*)
  - Innisfail, Queensland, Australia (*Af*)
  - Kisangani, Congo (*Af*)
  - Kuching, Malaysia (*Af*)
  - Paramaribo, Suriname (*Af*)
  - Quibdó, Colombia (*Af*)
  - Ratnapura, Sri Lanka (*Af*)

- Shah Alam, Malaysia (*Af*)
- Singapore (*Af*)
- West Palm Beach, Florida, United States (*Af*)
- Hilo, Hawaii, United States (*Af*)

Some of the places with this climate are indeed uniformly and monotonously wet throughout the year (e.g., the northwest Pacific coast of South and Central America, from Ecuador to Costa Rica; see, for instance, Andagoya, Colombia), but in many cases, the period of higher sun and longer days is distinctly wettest (as at Palembang, Indonesia) or the time of lower sun and shorter days may have more rain (as at Sitiawan, Malaysia).

(**Note**. The term aseasonal refers to the lack in the tropical zone of large differences in daylight hours and mean monthly (or daily) temperature throughout the year. Annual cyclic changes occur in the tropics, but not as predictably as those in the temperate zone, albeit unrelated to temperature, but to water availability whether as rain, mist, soil, or ground water. Plant response (e. g., phenology), animal (feeding, migration, reproduction, etc.), and human activities (plant sowing, harvesting, hunting, fishing, etc.) are tuned to this 'seasonality'. Indeed, in tropical South America and Central America, the 'rainy season' (and the 'high water season') is called *invierno* or *inverno*, though it could occur in the Northern Hemisphere summer; likewise, the 'dry season' (and 'low water season') is called *verano* or *verão*, and can occur in the Northern Hemisphere winter).

Tropical Monsoon Climate

This type of climate, most common in South Asia, results from the monsoon winds which change direction according to the seasons. This climate has a driest month (which nearly always occurs at or soon after the "winter" solstice for that side of the equator) with rainfall less than 60 mm, but more than 1/25 the total annual precipitation.

- Examples:
  - Abidjan, Ivory Coast (*Am*)
  - Cairns, Queensland, Australia (*Am*)
  - Chittagong, Bangladesh (*Am*)
  - Guanare, Portuguesa, Venezuela (*Am*)
  - Huế, Thừa Thiên–Huế, Vietnam (*Am*)
  - Baguio, Philippines (*Am*, bordering on *Cwb*)
  - Jakarta, Indonesia (*Am*)
  - Macapá, Amapá, Brazil (*Am*)
  - Miami, Florida, United States (*Am*)
  - Port Harcourt, Rivers State, Nigeria (*Am*)
  - Puerto Ayacucho, Amazonas, Venezuela (*Am*)
  - Sinop, Mato Grosso, Brazil (*Am*)
  - Yangon, Myanmar (*Am*)
  - Malé, Maldives (*Am*)

Tropical Wet and Dry or Savanna Climate

Aw climates have a pronounced dry season, with the driest month having precipitation less than 60 mm and less than 1/25 of the total annual precipitation.

- Examples:
  - Phnom Penh, Cambodia (*Aw*)
  - Barquisimeto, Venezuela (*Aw*)
  - Cartagena, Colombia (*Aw*)
  - Ciudad Guayana, Venezuela (*Aw*)
  - Dar es Salaam, Tanzania (*Aw*)
  - Darwin, Northern Territory, Australia (*Aw*)
  - Katherine, Northern Territory, Australia (*Aw*)
  - Sanya, Hainan, China (*Aw*)
  - Dili, East Timor (*Aw*)
  - Ho Chi Minh City, Vietnam (*Aw*)
  - Kaohsiung, Taiwan (*Aw*)
  - Kupang, Indonesia (*Aw*)
  - Kingston, Jamaica (*Aw*)
  - Lagos, Lagos State, Nigeria (*Aw*)
  - Medellín, Colombia (*Aw*)
  - Mumbai, Maharashtra, India (*Aw*)
  - Naples, Florida, United States (*Aw*)
  - Port-au-Prince, Haiti (*Aw*)
  - Rio de Janeiro, Rio de Janeiro, Brazil (*Aw*, bordering on *Am*)
  - Veracruz, Veracruz, Mexico (*Aw*)
  - Cancún, Mexico (*Aw*)
  - Guatemala City, Guatemala (*Aw*)
  - Dhaka, Bangladesh (*Aw*)
  - Panama City (*Aw*)

Most places that have this climate are found at the outer margins of the tropical zone from the low teens to the mid-20s latitudes, but occasionally an inner-tropical location (e.g., San Marcos, Antioquia, Colombia) also qualifies. Actually, the Caribbean coast, eastward from the Gulf of Urabá on the Colombia–Panamá border to the Orinoco River delta, on the Atlantic Ocean (about 4,000 km), have long dry periods (the extreme is the *BSh* climate (see below), characterized by very low, unreliable precipitation, present, for instance, in extensive areas in the Guajira, and Coro, western Venezuela, the northernmost peninsulas in South America, which receive <300 mm total annual precipitation, practically all in two or three months).

This condition extends to the Lesser Antilles and Greater Antilles forming the circum-Caribbean dry belt. The length and severity of the dry season diminishes inland (southward); at the latitude of the Amazon River—which flows eastward, just south of the equatorial line—the climate is *Af*. East from the Andes, between the dry, arid Caribbean and the ever-wet Amazon are the Orinoco River's llanos or savannas, from where this climate takes its name.

Sometimes *As* is used in place of *Aw* if the dry season occurs during the time of higher sun and longer days. This is the case in parts of Hawaii, northwestern Dominican Republic (Monte Cristi, Villa Vásquez, Luperón), East Africa (Mombasa, Kenya), Sri Lanka (Trincomalee), and Brazilian Northeastern Coast (Recife, João Pessoa and Natal), for instance. In most places that have tropical wet and dry climates, however, the dry season occurs during the time of lower sun and shorter days because of rain shadow effects during the 'high-sun' part of the year.

Group B: Dry (Arid and Semiarid) Climates

These climates are characterized by actual precipitation less than a threshold value set equal to the potential evapotranspiration. The threshold value (in millimeters) is determined as:

Multiply the average annual temperature in °C by 20, then add (a) 280 if 70% or more of the total precipitation is in the high-sun half of the year (April through September in the Northern Hemisphere, or October through March in the Southern), or (b) 140 if 30%–70% of the total precipitation is received during the applicable period, or (c) 0 if less than 30% of the total precipitation is so received.

According to the modified Köppen classification system used by modern climatologists, total precipitation in the warmest six months of the year is taken as reference instead of the total precipitation in the high-sun half of the year.

If the annual precipitation is less than 50% of this threshold, the classification is *BW* (arid: desert climate); if it is in the range of 50%–100% of the threshold, the classification is *BS* (semi-arid: steppe climate).

A third letter can be included to indicate temperature. Originally, *h* signified low-latitude climate (average annual temperature above 18 °C) while *k* signified middle-latitude climate (average annual temperature below 18 °C), but the more common practice today, especially in the United States, is to use *h* to mean the coldest month has an average temperature above 0 °C (32 °F), with *k* denoting that at least one month averages below 0 °C.

Desert areas situated along the west coasts of continents at tropical or near-tropical locations are characterized by cooler temperatures than encountered elsewhere at comparable latitudes (due to the nearby presence of cold ocean currents) and frequent fog and low clouds, despite the fact that these places rank among the driest on earth in terms of actual precipitation received. This climate is sometimes labelled *BWn*. The *BSn* category can be found in foggy coastal steppes.

- Arid climate examples:
    - Coober Pedy, Australia (*BWh*)

- Alice Springs, Australia (*BWh*)
- Baghdad, Iraq (*BWh*)
- Upington, Northern Cape South Africa (*BWh*)
- El Paso, Texas, United States (*BWh*)
- Phoenix, Arizona, United States (*BWh*)
- Death Valley, California, United States (*BWh*), location of the hottest air temperature ever recorded on Earth.
- Las Vegas, Nevada, United States (*BWh*)
- Hermosillo, Sonora, Mexico (*BWh*)
- Doha, Qatar (*BWh*)
- ‘Aziziya, Jafara, Libya (*BWh*)
- Mecca, Makkah Region, Saudi Arabia (*BWh*)
- Riyadh, Saudi Arabia (*BWh*)
- Kuwait City, Capital Governorate, Kuwait (*BWh*)
- Khartoum, Sudan (*BWh*)
- Dubai, United Arab Emirates (*BWh*)
- Turpan, Xinjiang, China (*BWk*)
- Nukus, Uzbekistan (*BWk*)
- Almería, Andalusia, Spain (*BWk*)
- Lima, Peru (*BWn*)
- Walvis Bay, Erongo Region, Namibia (*BWn*)

■ Semi-arid examples:

- Lahore, Punjab, Pakistan (*BSh*)
- Oranjestad, Aruba (*BSh*)
- Alicante, Spain (*BSh*)
- Ivanhoe, New South Wales, Australia (*BSh*)
- Piraeus, Greece (*BSh*)
- Bamako, Mali (*BSh*)
- Ouagadougou, Burkina Faso (*BSh*)
- Niamey, Niger (*BSh*)
- N’Djamena, Chad (*BSh*)
- Yuanjiang, Yunnan, China (*BSh*)
- Amman, Amman Governorate, Jordan (*BSh*)
- Ciudad Obregón, Sonora, Mexico (*BSh*)
- Navojoa, Sonora, Mexico (*BSh*)
- Baku, Azerbaijan (*BSk*)

- Yerevan, Armenia (*BSk*)
- Denver, Colorado, United States (*BSk*)
- Zaragoza, Spain (*BSk*)
- Zacatecas City, Zacatecas, Mexico (*BSk*)
- Tabriz, East Azerbaijan Province, Iran (*BSk*)
- Ulan-Ude, Russia (*BSk*)
- Lethbridge, Alberta, Canada (*BSk*)
- Brooks, Alberta, Canada (*BSk*)
- Shijiazhuang, Hebei, China (*BSk*)
- Lhasa, Tibet Autonomous Region, China (*BSk*)
- Ulaanbaatar, Mongolia (*BSk*)
- Comodoro Rivadavia, Chubut Province, Argentina (*BSk*)
- Sarmiento, Chubut, Argentina (*BSk*)

On occasion, a fourth letter is added to indicate if either the winter or summer half of the year is "wetter" than the other half. To qualify, the wettest month must have at least 60 mm (2.4 in) of average precipitation if all 12 months are above 18 °C (64 °F), or 30 mm (1.2 in) if not; plus at least 70% of the total precipitation must be in the same half of the year as the wettest month — but the letter used indicates when the ‹dry' season occurs, not the ‹wet' one. This would result in Khartoum, Sudan, being reckoned as *BWhw*; Niamey, Niger, as *BShw*; Alexandria, Egypt, as *BWhs*; Asbi'ah, Libya, as *BShs*; Ömnögovi Province, Mongolia, as *BWkw*; and Xining, Qinghai, China, as *BSkw* (*BWks* and *BSks* do not exist if 0 °C in the coldest month is recognized as the *h*/*k* boundary.) If the standards for neither *w* nor *s* are met, no fourth letter is added.

Group C: Temperate/Mesothermal Climates

In the Koppen climate system, temperate climates are defined as having an average temperature above -3 C (26.6 F) (or 0 °C in some versions, as noted previously) in their coldest month but below 18 C (64.4 F). The average temperature of -3 C roughly coincides with the equatorward limit of frozen ground and snowcover lasting for a month or more.

The second letter indicates the precipitation pattern—*w* indicates dry winters (driest winter month average precipitation less than one-tenth wettest summer month average precipitation. *s* indicates at least three times as much rain in the wettest month of winter as in the driest month of summer. *f* means significant precipitation in all seasons (neither above-mentioned set of conditions fulfilled).

The third letter indicates the degree of summer heat—*a* indicates warmest month average temperature above 22 °C (72 °F) while *b* indicates warmest month averaging below 22 °C.

Mediterranean Climates

These climates usually occur on the western sides of continents between the latitudes of 30° and 45°. These climates are in the polar front region in winter, and thus have moderate temperatures and changeable, rainy weather. Summers are hot and dry, due to the domination of the subtropical high pressure systems, except in the immediate coastal areas, where summers are milder due to the nearby presence of cold ocean currents that may bring fog but prevent rain.

- Examples:
  - Beirut, Lebanon (*Csa*)
  - Latakia, Syria (*Csa*)
  - Tel Aviv, Israel (*Csa*)
  - Los Angeles, California, United States (*Csa* bordering on *Csb*, *BSh*, and *BSk*)
  - San Bernardino, California, United States (*Csa*)
  - Sacramento, California, United States (*Csa*)
  - Medford, Oregon, United States (*Csa*)
  - Walla Walla, Washington, United States (*Csa*)
  - Perth, Australia (*Csa*)
  - Adelaide, Australia (*Csa*)
  - Tangier, Morocco(*Csa*)
  - Casablanca, Morocco (*Csa*)
  - Rome, Italy (*Csa*)
  - Seville, Spain (*Csa*)
  - Palma de Mallorca, Spain (*Csa*)
  - Madrid, Spain (*Csa*)
  - Marseille, France (*Csa*)
  - Nice, France (*Csa*)
  - Athens, Greece (*Csa*)
  - Antalya, Turkey (*Csa*)
  - Tashkent, Uzbekistan (*Csa*)
  - Dushanbe, Tajikistan (*Csa*)
  - Algiers, Algeria (*Csa*)
  - Tunis, Tunisia (*Csa*)
- Dry-summer sometimes extends to additional areas not typically associated with a typical Mediterranean climate, as their warmest month mean doesn't reach 22 °C (71.6 °F), they are classified as (*Csb*). Some of these areas would border the Oceanic climate (*Cfb*), except their dry-summer patterns meet Köppen's *Cs* minimum thresholds.

- Examples include
    - Santiago, Chile (*Csb*)
    - Essaouira, Morocco (*Csb*)
    - Porto, Portugal (*Csb*)
    - Potenza, Italy (*Csb*)
    - San Francisco, California, United States (*Csb*)
    - Eureka, California, United States (*Csb*)
    - Newport, Oregon, United States (*Csb*)
    - Seattle, Washington, United States (*Csb*)
    - Victoria, British Columbia, Canada (*Csb*)
    - Cape Town, South Africa (*Csb*)
    - Kingston SE, Australia (*Csb*)
    - Albany, Western Australia, Australia (*Csb*)
    - Flinders Island, Australia (*Csb*)

■ Cold-summer Mediterranean climates (*Csc*) exist in high-elevation areas adjacent to coastal *Csb* climate areas, where the strong maritime influence prevents the average winter monthly temperature from dropping below 0 °C. This climate is rare and is predominantly found in climate fringes and isolated areas of the Cascades and Andes Mountains, as the dry-summer climate extends further poleward in the Americas than elsewhere. Rare instances of this climate can be found in some coastal locations in the North Atlantic and at high altitudes in Hawaii.

- Examples:
    - Balmaceda, Chile (*Csc*)
    - Haleakalā Summit, Hawaii (*Csc*)

Humid Subtropical Climates

These climates usually occur on the eastern coasts and eastern sides of continents, usually in the high 20s and 30s latitudes. Unlike the dry summer Mediterranean climates, humid subtropical climates have a warm and wet flow from the tropics that creates warm and moist conditions in the summer months. As such, summer (not winter as is the case in Mediterranean climates) is often the wettest season.

The flow out of the subtropical highs and the summer monsoon creates a southerly flow from the tropics that brings warm and moist air to the lower east sides of continents. This flow is often what brings the frequent but short-lived summer thundershowers so typical of the more southerly subtropical climates like the far southern United States, southern China and Japan.

- Examples:
  - Buenos Aires, Argentina (*Cfa*)
  - Porto Alegre, Brazil (*Cfa*)
  - São Paulo, Brazil (*Cfa*)
  - Brisbane, Australia (*Cfa*)
  - Gold Coast, Queensland, Australia (*Cfa*)
  - Barcelona, Spain (*Cfa*)
  - Girona, Spain (*Cfa*)
  - Huesca, Spain (*Cfa*)
  - Rasht, Gilan Province, Iran (*Cfa*)
  - New Orleans, Louisiana, United States (*Cfa*)
  - Orlando, Florida, United States (*Cfa*)
  - Charleston, South Carolina, United States (*Cfa*)
  - Nashville, Tennessee, United States (*Cfa*)
  - Washington D.C., United States (*Cfa*)
  - Houston, Texas, United States (*Cfa*)
  - Atlanta, Georgia, United States (*Cfa*)
  - Astara, Azerbaijan (*Cfa*)
  - Horta, Azores (*Cfa*)
  - Rize, Turkey (*Cfa*)
  - Sochi, Russia (*Cfa*)
  - Constan□a, Romania (*Cfa*, bordering on *Cfb*)
  - Kutaisi, Georgia (*Cfa*)
  - Srinagar, India (*Cfa*)
  - Shanghai, China (*Cfa*)
  - Nanjing, China (*Cfa*)
  - Taipei, Taiwan (*Cfa*)
  - Tokyo, Japan (*Cfa*)
  - Osaka, Japan (*Cfa*)
  - Milan, Italy (*Cfa*)
  - Turin, Italy (*Cfa*)
  - Venice, Italy (*Cfa*)
  - Tirana, Albania (*Cfa*)
  - Durban, South Africa (*Cfa*)

New York City is on the borderline between this climate and the humid continental (*Dfa*) climate, the line passing through the city.

- Subtropical-Dry Winter (*Cwa*) is monsoonal influenced, having the classic dry winter/wet summer pattern associated with tropical monsoonal climates.
    - Examples include:
        - Córdoba, Argentina (*Cwa*)
        - Santiago del Estero, Argentina (*Cwa*)
        - Mackay, Queensland, Australia (*Cwa*)
        - Islamabad, Pakistan (*Cwa*)
        - New Delhi, India (*Cwa*)
        - Haikou, China (*Cwa*)
        - Zhengzhou, China (*Cwa*)
        - Xi'an, China (*Cwa*)
        - Chengdu, Sichuan China (*Cfa*)
        - Hong Kong (*Cwa*)
        - Kathmandu, Nepal (*Cwa*)
        - Imphal, India (*Cwa*)
        - Taunggyi, Myanmar (*Cwa*)
        - Hanoi, Vietnam (*Cwa*)
        - Guadalajara, Mexico (*Cwa*)

Oceanic and Subtropical Highland Climates

Cfb climates usually occur in the higher middle latitudes on the western sides of continents between the latitudes of 40° and 60°; they are typically situated immediately poleward of the Mediterranean climates, although in Australia and extreme southern Africa this climate is found immediately poleward of temperate climates, and at a somewhat lower latitude. In western Europe, this climate occurs in coastal areas up to 63°N in Norway.

These climates are dominated all year round by the polar front, leading to changeable, often overcast weather. Summers are cool due to cool ocean currents, but winters are milder than other climates in similar latitudes, but usually very cloudy. *Cfb* climates are also encountered at high elevations in certain subtropical and tropical areas, where the climate would be that of a subtropical/tropical rain forest if not for the altitude. These climates are called "highlands".

- Examples:
    - Paris, France (*Cfb*)
    - Dunkirk, France (*Cfb*)
    - Bergen, Norway (*Cfb*)
    - Berlin, Germany (*Cfb*)
    - London, England, United Kingdom (*Cfb*)
    - Southampton, England, United Kingdom (*Cfb*)
    - Dublin, Ireland (*Cfb*)
    - Glasgow, Scotland (*Cfb*)
    - Edinburgh, Scotland (*Cfb*)
    - Shetland, Scotland (*Cfb*, bordering on *Cfc*)
    - Amsterdam, Netherlands (*Cfb*)
    - Bilbao, Spain (*Cfb*)
    - Santiago de Compostela, Spain (*Cfb*)
    - Zonguldak, Turkey (*Cfb*)
    - Auckland, North Island, New Zealand (*Cfb*)
    - Mar del Plata, Buenos Aires Province, Argentina (*Cfb*)
    - Valdivia, Los Ríos Region, Chile (*Cfb*)
    - Quito, Pichincha Province, Ecuador (*Cfb*)
    - Bogotá, Colombia (*Cfb*)
    - São Joaquim, Brazil (*Cfb*)
    - Campos do Jordão, Brazil (*Cfb*)
    - Prince Rupert, British Columbia, Canada (*Cfb*)
    - Vancouver, British Columbia, Canada (*Cfb*, bordering on *Csb*)
    - Forks, Washington, United States (*Cfb*)
    - Ketchikan, Alaska, United States (*Cfb*)
    - Block Island, Rhode Island, United States (*Cfb*)
    - Chatham, Massachusetts, United States (*Cfb*, using the −3 °C, 27 °F, isotherm)
    - Boone, North Carolina, United States (*Cfb*)
    - George, Western Cape, South Africa (*Cfb*)
    - Port Elizabeth, South Africa (*Cfb*)
    - Melbourne, Victoria, Australia (*Cfb*)
    - Hobart, Tasmania, Australia (*Cfb*)
    - Tristan da Cunha in the southern Atlantic Ocean (*Cfb*)
    - Île Amsterdam in the southern Indian Ocean (*Cfb*)

    - Brinchang, Malaysia (*Cfb*)
    - Brussels, Belgium (*Cfb*)
    - Andorra la Vella, Andorra (*Cfb*)
    - Constanza, Dominican Republic (*Cfb*)
    - Skagen, Denmark (*Cfb*, bordering on *Dfb*)
    - Copenhagen, Denmark (*Cfb*, bordering on *Dfb*)
    - Kaliningrad, Russia (*Cfb*, bordering on *Dfb*)
    - Zürich, Switzerland (*Cfb*, bordering on *Dfb*)
    - Budapest, Hungary (*Cfb*, bordering on *Dfb*, *Dfa*, and *Cfa*)
    - Collinsvale, Tasmania, Australia (*Cfb* bordering on *Dfb*)
- Subpolar oceanic climates (*Cfc*) occur poleward of or at higher elevations than the maritime temperate climates, and are mostly confined either to narrow coastal strips on the western poleward margins of the continents, or, especially in the Northern Hemisphere, to islands off such coasts. They occur in both hemispheres, most often at latitudes from 60° north and south to 70° north and south.
- Examples:
    - Grouse Mountain, North Vancouver, British Columbia (*Cfc*)
    - Reykjavík, Iceland (*Cfc*)
    - Unalaska, Alaska, United States (*Cfc*)
    - Tórshavn, Faroe Islands (*Cfc*)
    - Mount Read, Tasmania, Australia (*Cfc*, borders on ET)
    - Auckland Islands, New Zealand (*Cfc*)
    - Punta Arenas, Chile (*Cfc*)

Highland Climate with Dry Winters

This is a type of climate mainly found in highlands inside the tropics of Mexico, Peru, Bolivia, Madagascar, Zambia, Zimbabwe and South Africa, but it is also found in east-central Spain or central Argentina and areas outside of the tropics. Winters are noticeable and dry, and summers can be very rainy. In the tropics, the rainy season is provoked by the tropical air masses and the dry winters by subtropical high pressure.

- Examples:
    - Da Lat, Vietnam (*Cwb*)
    - Sa Pa, Vietnam (*Cwb*)
    - Mokhotlong, Lesotho (*Cwb*)
    - Nairobi, Kenya (*Cwb*)
    - Sagada, Philippines (*Cwb*)
    - Kunming, China (*Cwb*)
    - Fraijanes, Guatemala (*Cwb*)

- Addis Ababa, Ethiopia (*Cwb*)
- El Alto, Bolivia (*Cwc*)

Group D: Continental/Microthermal Climates

These climates have an average temperature above 10 °C (50 °F) in their warmest months, and a coldest month average below −3 °C (or 0 °C in some versions, as noted previously).

The snowy city of Sapporo

These usually occur in the interiors of continents and on their upper east coasts, normally north of 40°N. In the Southern Hemisphere, group D climates are extremely rare due to the smaller land masses in the middle latitudes and the almost complete absence of land at 40–60°S, existing only in some highland locations.

Hot Summer Continental Climates

Dfa climates usually occur in the high 30s and low 40s latitudes, with a qualifying average temperature in the warmest month of >22 °C/72 °F. In Europe, these climates tend to be much drier than in North America. In eastern Asia, *Dwa* climates extend further south due to the influence of the Siberian high pressure system, which also causes winters there to be dry, and summers can be very wet because of monsoon circulation. *Dsa* exists at higher elevations adjacent to areas with hot summer Mediterranean (*Csa*) climates.

- Examples:
  - Almaty, Kazakhstan (*Dfa*)
  - Oral, Kazakhstan (*Dfa*)

- Sapporo, Japan (*Dfa*, bordering on *Dfb*)
- Aomori, Aomori Prefecture, Japan (*Dfa*)
- Nagano, Nagano Prefecture, Japan (*Dfa*)
- Chicago, Illinois, United States (*Dfa*)
- Cleveland, Ohio, United States (*Dfa*)
- Columbus, Ohio, United States (*Dfa*)
- Sioux Falls, South Dakota, United States (*Dfa*)
- Omaha, Nebraska, United States (*Dfa*)
- Minneapolis, Minnesota, United States (*Dfa*)
- Detroit, Michigan, United States (*Dfa*)
- Salt Lake City, Utah, United States (*Dfa*)
- Bucharest, Romania (*Dfa*)
- Rostov-on-Don, Russia (*Dfa*)
- Pyongyang, North Korea (*Dwa*)
- Beijing, China (*Dwa*)
- Harbin, China (*Dwa*)
- Tianjin, China (*Dwa*)
- Changchun, China (*Dwa*)
- Shenyang, China (*Dwa*)
- Seoul, South Korea (*Dwa*)

New York City is on the borderline between this climate and the humid subtropical (*Cfa*) climate, the line passing through the city.

- *Dsa* exists only at higher elevations adjacent to areas with hot summer Mediterranean (*Csa*) climates.
- Examples include:
  - Sanandaj, Kurdistan Province, Iran (*Dsa*)
  - Saqqez, Kurdistan Province, Iran (*Dsa*)
  - Arak, Markazi Province, Iran (*Dsa*)
  - Hakkâri, Turkey (*Dsa*)
  - Muş, Turkey (*Dsa*)
  - Bishkek, Kyrgyzstan (*Dsa*)
  - Urzhar, Kazakhstan (*Dsa*, bordering on *Dsb*)
  - Cambridge, Idaho, United States (*Dsa*)

Warm Summer Continental or Hemiboreal Climates

Dfb and *Dwb* climates are immediately north of hot summer continental climates, generally in the high 40s and low 50s latitudes in North America and Asia, and also extending to higher latitudes in central and eastern Europe and Russia, between the maritime temperate and continental subarctic climates, where it extends up to 65 degrees latitude in places.

- Examples:
    - Helsinki, Finland (*Dfb*)
    - Svolvær, Lofoten, Nordland, Norway (*Dfb*, bordering on *Cfb*, *Cfc*, and *Dfc*)
    - Kiev, Ukraine (*Dfb*)
    - Moscow, Russia (*Dfb*)
    - Saint Petersburg, Russia (*Dfb*)
    - Stockholm, Sweden (*Dfb*)
    - Oslo, Norway (*Dfb*)
    - Lillehammer, Norway (*Dfb*)
    - Portland, Maine, United States (*Dfb*)
    - Montpelier, Vermont, United States (*Dfb*)
    - Binghamton, New York, United States (*Dfb*)
    - Worcester, Massachusetts, United States (*Dfb*)
    - Youngstown, Ohio, United States (*Dfb*)
    - Aspen, Colorado, United States (*Dfb*)
    - Juneau, Alaska, United States (*Dfb*, bordering on *Cfb*, *Cfc*, and *Dfc*)
    - Calgary, Alberta, Canada (*Dfb*)
    - Saskatoon, Saskatchewan, Canada (*Dfb*)
    - Winnipeg, Manitoba, Canada (*Dfb*)
    - Ottawa, Ontario, Canada (*Dfb*)
    - Toronto Islands, Ontario, Canada (*Dfb*)
    - London, Ontario, Canada (*Dfb*)
    - Edmonton, Alberta, Canada (*Dfb*)
    - Montreal, Quebec, Canada (*Dfb*)
    - Quebec City, Quebec, Canada (*Dfb*)
    - Moncton, New Brunswick, Canada (*Dfb*)
    - Halifax, Nova Scotia, Canada (*Dfb*)
    - Charlottetown, Prince Edward Island, Canada (*Dfb*)
    - St. John's, Newfoundland and Labrador, Canada (*Dfb*)
    - Erzurum, Turkey (*Dfb*)
    - Ardahan, Turkey (*Dfb*)
    - Warsaw, Poland (*Dfb*)
    - Prague, Czech Republic (*Dfb*)
    - Minsk, Belarus (*Dfb*)

- Kiandra, New South Wales, Australia (*Dfb*)
- Cabramurra, New South Wales, Australia (*Dfb*)
- Kushiro, Hokkaido, Japan (*Dfb*)
- Karaganda, Kazakhstan (*Dfb*)
- Dickinson, North Dakota, United States (*Dwb*)
- Heihe, China (*Dwb*)
- Vladivostok, Russia (*Dwb*)
- Irkutsk, Russia (*Dwb*)
- Baruunturuun, Mongolia (*Dwb*)
- Pyeongchang County, South Korea (*Dwb*)

- Dsb arises from the same scenario as Dsa, but at even higher altitudes or latitudes, and chiefly in North America, since the Mediterranean climates extend further poleward than in Eurasia.
- Examples include:
    - Sivas, Turkey (*Dsb*)
    - Yüksekova, Turkey (*Dsb*)
    - Roghun, Tajikistan (*Dsb*)
    - Arslanbob, Kyrgyzstan (*Dsb*)
    - Flagstaff, Arizona, United States (*Dsb*)
    - South Lake Tahoe, California, United States (*Dsb*)
    - Bridgeport, California, United States (*Dsb*)
    - Wallace, Idaho, United States (*Dsb*)
    - Alexandrovsk-Sakhalinsky, Russia (*Dsb*, bordering on *Dfb*)
    - Dras, India (*Dsb*)

Subarctic or Boreal Climates

Dfc and *Dwc* climates occur poleward of the other group D climates, generally in the 50s and low 60s North latitudes. In some places, it extends northward to beyond 70°N latitude.

- Examples:
    - Murmansk, Murmansk Oblast, Russia (*Dfc*)
    - Arkhangelsk, Russia (*Dfc*)
    - Aldan, Russia (*Dfc*)
    - Yellowknife, Northwest Territories, Canada (*Dfc*)
    - Rankin Inlet, Nunavut, Canada (*Dfc*)
    - Kuujjuaq, Quebec, Canada (*Dfc*)
    - Matagami, Quebec, Canada (*Dfc*)
    - Whitehorse, Yukon, Canada (*Dfc*)

- Churchill, Manitoba, Canada (*Dfc*)
- Labrador City, Newfoundland and Labrador, Canada (*Dfc*)
- Nain, Newfoundland and Labrador, Canada (*Dfc*)
- Fraser, Colorado, United States (*Dfc*)
- Fairbanks, Alaska, United States (*Dfc*)
- Anchorage, Alaska, United States (*Dfc*)
- Nome, Alaska, United States (*Dfc*)
- Sütoluk, Ardahan Province, Turkey (*Dfc*)
- Mount Buller, Victoria, Australia (*Dfc*)
- Charlotte Pass, New South Wales, Australia (*Dfc*)
- St. Moritz, Grisons, Switzerland (*Dfc*)
- Tromsø, Norway (*Dfc*)
- Tignes, Savoie, France (*Dfc*)
- Homer, Alaska, United States (*Dsc*)
- Bodie, California, United States (*Dsc*)
- Brian Head, Utah, United States (*Dsc*)
- Bohemia Mountain, Oregon, United States (*Dsc*)
- Spirit Lake, Washington, United States (*Dsc*), previous to 1980 eruption of Mount St. Helens
- Khandud, Badakhshan Province, Afghanistan (*Dsc*)
- Mohe County, Heilongjiang, China (*Dwc*)
- Mörön, Mongolia (*Dwc*)
- Yushu City, Qinghai, China (*Dwc*)

■ Places with this climate (*Dfd,Dwd,Dsd*) have severe winters, with the temperature in their coldest month lower than −38 °C. These climates occur only in eastern Siberia. The names of some of the places with this climate have become veritable synonyms for extreme, severe winter cold.

- Examples:
    - Yakutsk, Sakha Republic, Russia (*Dfd*)
    - Verkhoyansk, Sakha Republic, Russia (*Dfd*)
    - Seymchan, Magadan Oblast, Russia (*Dfd*)
    - Oymyakon, Sakha Republic, Russia (*Dwd*)

Group E: Polar Climates

In the Koppen climate system, Polar climates are defined as the warmest temperature of any month is below 10 C (50 F). Polar climates are further divided into two types, Tundra climates and Icecap Climates:

- **Tundra climate** (*ET* and *ETf*): Warmest month has an average temperature between 0 and 10 °C. These climates occur on the northern edges of the North American and Eurasian land masses (generally north of 70 °N although it may be found farther south depending on local conditions), and on nearby islands. ET climates are also found on some islands near the Antarctic Convergence, and at high elevations outside the polar regions, above the tree line.
- Examples:
  - Macquarie Island (*ET*)
  - Crozet Islands (*ET*)
  - Campbell Island, New Zealand (*ET*)
  - Kerguelen Islands (*ET*)
  - Prince Edward Islands (*ET*)
  - Stanley, Falkland Islands (*ET*), borders subpolar oceanic (*Cfc*)
  - Ushuaia, Argentina (*ET*), (borders on *Cfc*)
  - La Rinconada, Peru (*ET*)
  - Mount Wellington, Australia (*ET*)
  - Nagqu, Tibet, China (*ET*)
  - Letseng diamond mine, Lesotho (*ET*, bordering on *Cfc* and *Dfc*)

These climates are a colder and more continental variants of tundra. They would have characteristics of the ice cap climate, but still manage to see temperatures above 0 °C (32 °F):

- Nanortalik, Greenland (*ETf*)
- Mount Fuji, Japan (*ETf*)
- Murghab, Tajikistan (*ETf*)
- Mount Washington, New Hampshire, United States (*ETf*)
- Eureka, Nunavut, Canada (*ETf*)
- Resolute, Nunavut, Canada (*ETf*)
- Iqaluit, Nunavut, Canada (*ETf*)
- Alert, Nunavut, Canada (*ETf*)
- Inukjuak, Quebec, Canada (*ETf*)
- Nuuk, Greenland (*ETf*)
- Svalbard (*ETf*)
- Mys Shmidta, Russia (*ETf*)
- Dikson Island, Russia (*ETf*)
- Nord, Greenland (*ETf*)
- Esperanza Base, Antarctica (*ETf*)

- North Pole (*ETf*, bordering on *EF*)

Ice cap climate (*EF*): This climate is dominant in Antarctica and inner Greenland, but also occurs at extremely high altitudes on mountains, above even tundra. All twelve months have average temperatures below 0 °C (32 °F).

- Examples:
  - Mount Rainier, Washington, United States (*EF*)
  - Mount Ararat, Turkey (*EF*)
  - Mount Everest, China/Nepal (*EF*)
  - Summit Camp, Greenland (*EF*)
  - Scott Base, Antarctica (*EF*)
  - Vostok Station, Antarctica (*EF*), location of the lowest air temperature ever recorded on Earth.
  - McMurdo Station, Antarctica (*EF*)
  - Byrd Station, Antarctica (*EF*)

Occasionally, a third, lower-case letter is added to ET climates (distinguishing between ETf, ETs, and ETw), if either the summer or winter is clearly drier than the other half of the year. When the option to include this letter is exercised, the same standards that are used for Groups C and D apply, with the additional requirement that the wettest month must have an average of at least 30 mm precipitation (Group E climates can be as dry or even drier than Group B climates based on actual precipitation received, but their rate of evaporation is much lower). Seasonal precipitation letters are almost never attached to EF climates, mainly due to the difficulty in distinguishing between falling and blowing snow, as snow is the sole source of moisture in these climates.

Occasionally, a third, lower-case letter is added to *ET* climates (distinguishing between *ETf*, *ETs*, and *ETw*), if either the summer or winter is clearly drier than the other half of the year. When the option to include this letter is exercised, the same standards that are used for Groups C and D apply, with the additional requirement that the wettest month must have an average of at least 30 mm precipitation (Group E climates can be as dry or even drier than Group B climates based on actual precipitation received, but their rate of evaporation is much lower). Seasonal precipitation letters are almost never attached to *EF* climates, mainly due to the difficulty in distinguishing between falling and blowing snow, as snow is the sole source of moisture in these climates.

Ecological Significance

The Köppen climate classification is based on the empirical relationship between climate and vegetation. This classification provides an efficient way to describe climatic conditions defined by temperature and precipitation and their seasonality with a single metric. Because climatic conditions identified by the Köppen classification are ecologically relevant, it has been widely used to map geographic distribution of long term climate and associated ecosystem conditions.

Over the recent years, there has been an increasing interest in using the classification to identify changes in climate and potential changes in vegetation over time. The most important ecological significance of the Köppen climate classification is that it helps to predict the dominant vegetation type based on the climatic data and vice versa.

In 2015, a pair of Chinese scholars published analysis of climate classifications between 1950 and 2010, finding that more than 5% of all land area worldwide had moved from wetter and colder classifications to drier and hotter classifications.

Trewartha Climate Classification Scheme

The Trewartha climate classification is a climate classification system published by American geographer Glenn Thomas Trewartha in 1966, and updated in 1980. It is a modified version of the 1899 Köppen system, created to answer some of the deficiencies of the Köppen system. The Trewartha system attempts to redefine the middle latitudes to be closer to vegetation zoning and genetic climate systems. It was considered a more true or "real world" reflection of the global climate.

For example, under the standard Köppen system, in the United States, western Washington and Oregon are classed into the same climate zone as southern California, even though the two regions have strikingly different weather and vegetation. Under the old Köppen system cool oceanic climates like that of London or Seattle were classed in the same zone as hot subtropical cities like Savannah, GA or Brisbane, Australia. In the United States, locations in the Midwest like Colorado and Iowa which have long, severe winter climates where plants are completely dormant, were classed into the same climate zone as Louisiana or northern Florida which have mild winters and a green winter landscape.

Other Maps

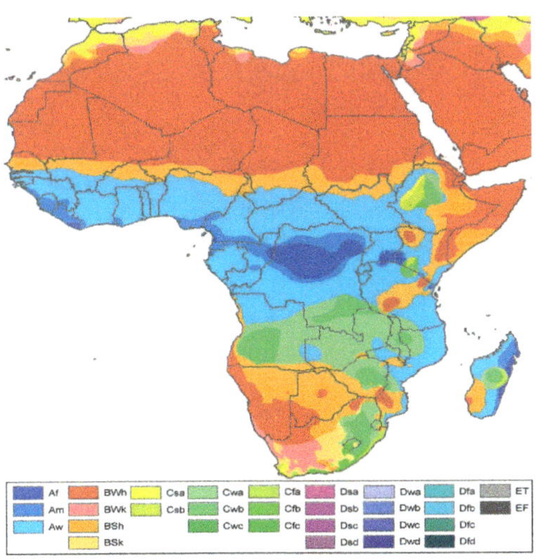

Köppen map of Africa

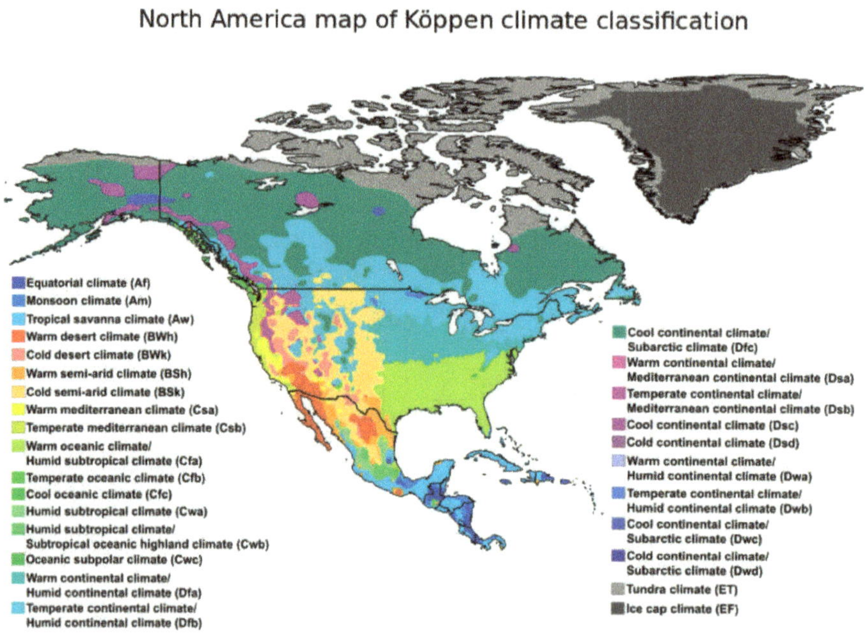

Köppen map of North America

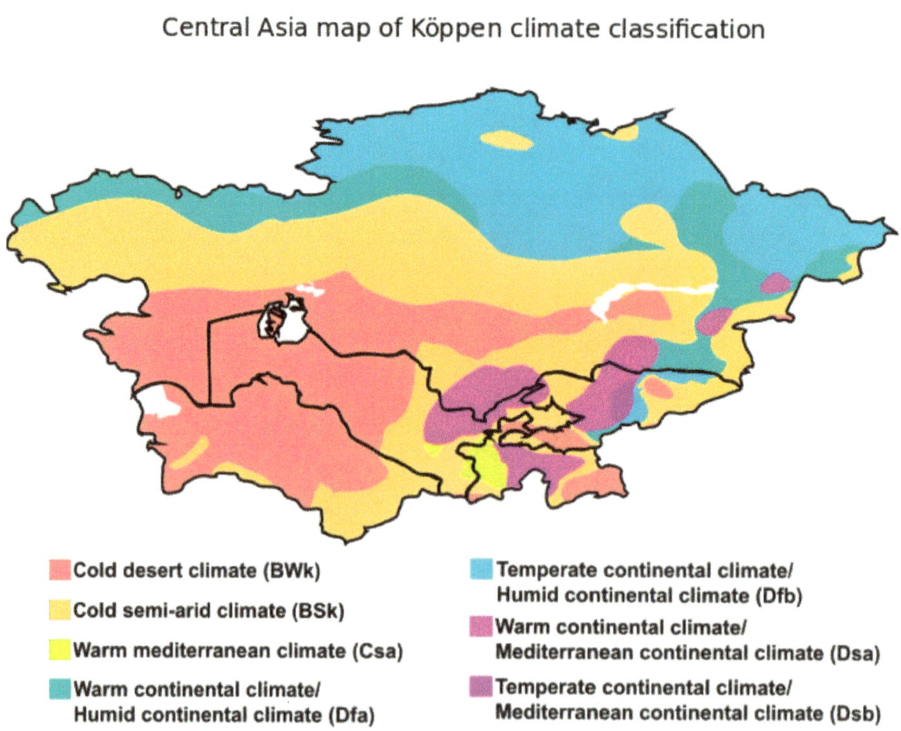

Köppen map of Central Asia

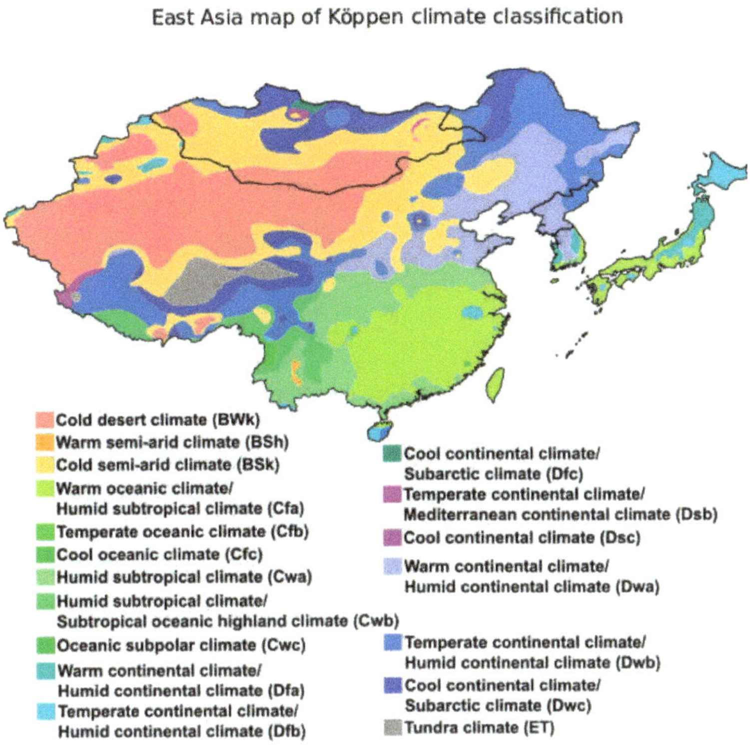

Köppen map of East Asia

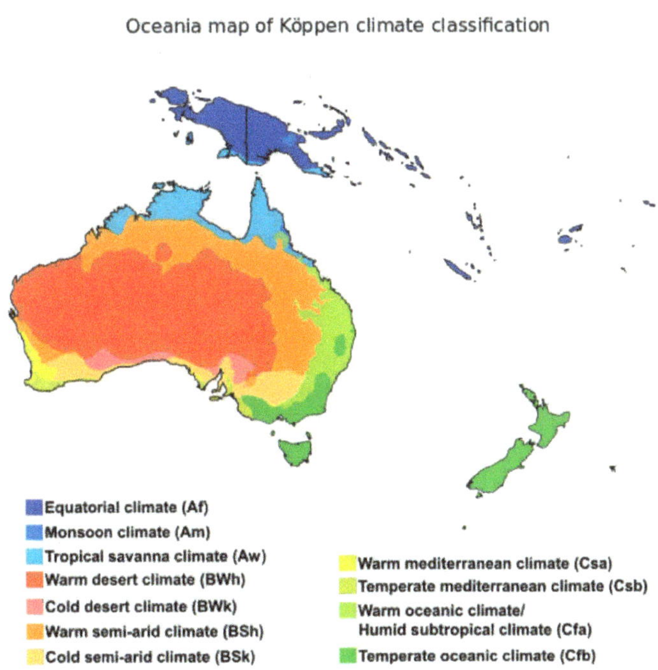

Köppen map of Australia/Oceania

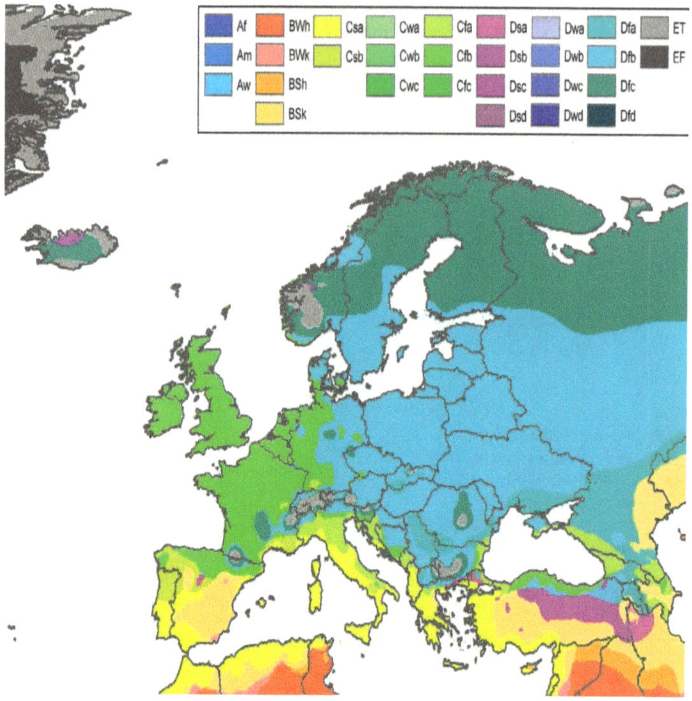

Köppen map of Europe

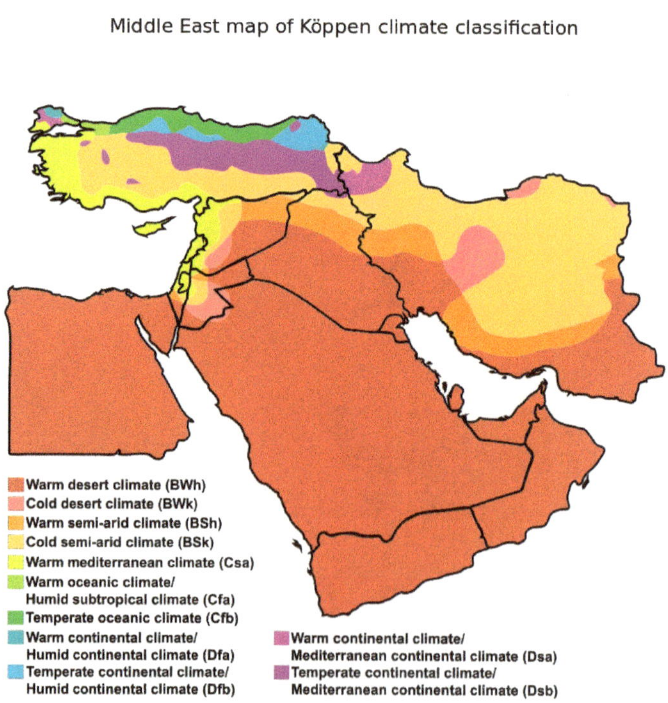

Köppen map of the Middle East

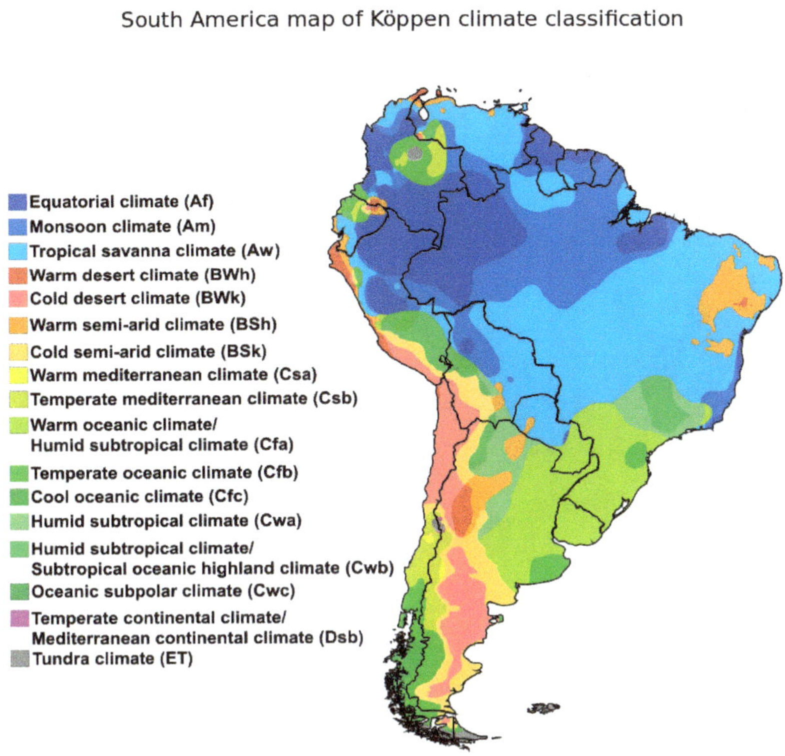

Köppen map of South America

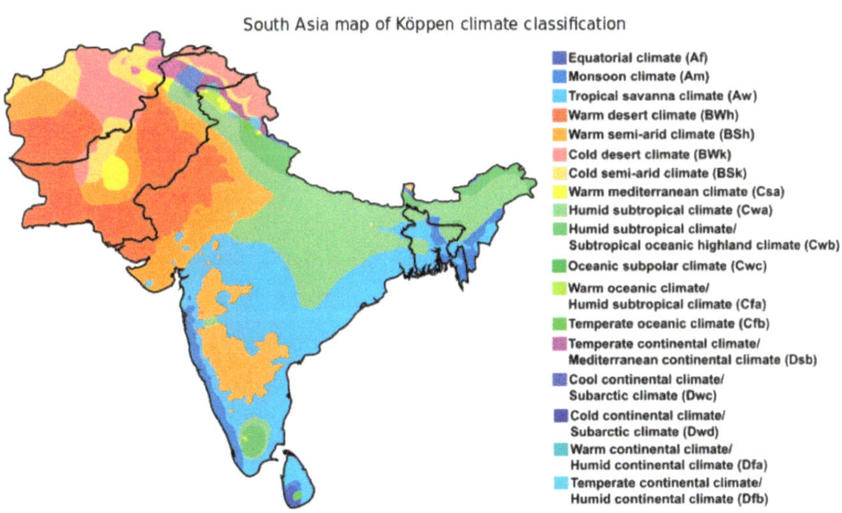

Köppen map of South Asia

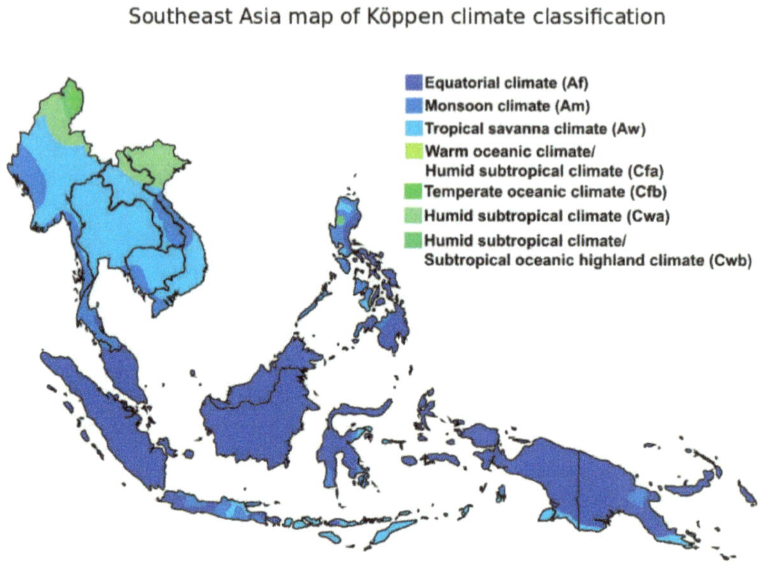

Köppen map of Southeast Asia

REFERENCES

1. Karger, Dirk Nikolaus; Conrad, Olaf; Böhner, Jürgen; Kawohl, Tobias; Kreft, Holger; Soria-Auza, Rodrigo W.; Zimmermann, Niklaus; Linder, H. Peter; Michael, Kessler (2016-07-01). "Climatologies at high resolution for the Earth land surface areas". arXiv:1607.00217 [physics.ao-ph].
2. "Precipitation". Glossary of Meteorology. American Meteorological Society. 2009. Retrieved 2009-01-02.
3. Scott Sistek (December 26, 2015). "What's the difference between 'rain' and 'showers'?". KOMO-TV. Retrieved January 18, 2016.
4. Chowdhury's Guide to Planet Earth (2005). "The Water Cycle". WestEd. Retrieved 2006-10-24.
5. Jim Lochner (1998). "Ask an Astrophysicist". NASA Goddard Space Flight Center. Retrieved 2009-01-16.
6. Emmanouil N. Anagnostou (2004). "A convective/stratiform precipitation classification algorithm for volume scanning weather radar observations". Meteorological Applications. Cambridge University Press. 11 (4): 291–300. Bibcode:2004MeApp..11..291A. doi:10.1017/S1350482704001409.
7. A.J. Dore, M. Mousavi-Baygi, R.I. Smith, J. Hall, D. Fowler and T.W. Choularton (June 2006). "A model of annual orographic precipitation and acid deposition and its application to Snowdonia". Atmospheric Environment. 40 (18): 3316–3326. Bibcode:2006AtmEn..40.3316D. doi:10.1016/j.atmosenv.2006.01.043.
8. Robert Penrose Pearce (2002). Meteorology at the Millennium. Academic Press. p. 66. ISBN 978-0-12-548035-2.ss

9. Robert A. Houze, Jr. (1994-06-28). Cloud Dynamics. Academic Press. p. 348. ISBN 0080502105.
10. Jan Jackson (2008). "All About Mixed Winter Precipitation". National Weather Service. Retrieved 2009-02-07.
11. Glossary of Meteorology (June 2000). "Dewpoint". American Meteorological Society. Retrieved 2011-01-31.
12. FMI (2007). "Fog And Stratus - Meteorological Physical Background". Zentralanstalt für Meteorologie und Geodynamik. Retrieved 2009-02-07.
13. Glossary of Meteorology (2009). "Adiabatic Process". American Meteorological Society. Retrieved 2008-12-27.
14. TE Technology, Inc (2009). "Peltier Cold Plate". Retrieved 2008-12-27.
15. Glossary of Meteorology (2009). "Radiational cooling". American Meteorological Society. Retrieved 2008-12-27.
16. Robert Fovell (2004). "Approaches to saturation" (PDF). University of California in Los Angeles. Retrieved 2009-02-07.
17. National Weather Service Office, Spokane, Washington (2009). "Virga and Dry Thunderstorms". Retrieved 2009-01-02.
18. Bart van den Hurk & Eleanor Blyth (2008). "Global maps of Local Land-Atmosphere coupling" (PDF). KNMI. Retrieved 2009-01-02.
19. H. Edward Reiley; Carroll L. Shry (2002). Introductory horticulture. Cengage Learning. p. 40. ISBN 978-0-7668-1567-4. Retrieved 2011-01-31.
20. National Weather Service JetStream (2008). "Air Masses". Retrieved 2009-01-02.
21. Michael Pidwirny (2008). "CHAPTER 8: Introduction to the Hydrosphere (e). Cloud Formation Processes". Physical Geography. Retrieved 2009-01-01.
22. Paul Sirvatka (2003). "Cloud Physics: Collision/Coalescence; The Bergeron Process". College of DuPage. Retrieved 2009-01-01.
23. United States Geological Survey (2009). "Are raindrops tear shaped?". United States Department of the Interior. Retrieved 2008-12-27.
24. . S. 0guntoyinbo and F. 0. Akintola (1983). "Rainstorm characteristics affecting water availability for agriculture" (PDF). IAHS Publication Number 140. Retrieved 2008-12-27.
25. Robert A. Houze Jr (1997). "Stratiform Precipitation in Regions of Convection: A Meteorological Paradox?". Bulletin of the American Meteorological Society. 78 (10): 2179–2196. Bibcode:1997BAMS...78.2179H. doi:10.1175/1520-0477(1997)078<2179:SPIROC>2.0.CO;2.
26. Norman W. Junker (2008). "An ingredients based methodology for forecasting precipitation associated with MCS's". Hydrometeorological Prediction Center. Retrieved 2009-02-07.
27. Alaska Air Flight Service Station (2007-04-10). "SA-METAR". Federal Aviation Administration via the Internet Wayback Machine. Archived from the original on 2008-05-01. Retrieved 2009-08-29.

28. “Hail (glossary entry)”. National Oceanic and Atmospheric Administration’s National Weather Service. Retrieved 2007-03-20.
29. Weatherquestions.com. “What causes ice pellets (sleet)?”. Retrieved 2007-12-08.
30. Glossary of Meteorology (2009). “Hail”. American Meteorological Society. Retrieved 2009-07-15.
31. Ryan Jewell & Julian Brimelow (2004-08-17). “P9.5 Evaluation of an Alberta Hail Growth Model Using Severe Hail Proximity Soundings in the United States” (PDF). Retrieved 2009-07-15.
32. National Severe Storms Laboratory (2007-04-23). “Aggregate hailstone”. National Oceanic and Atmospheric Administration. Retrieved 2009-07-15.
33. Julian C. Brimelow; Gerhard W. Reuter & Eugene R. Poolman (October 2002). “Modeling Maximum Hail Size in Alberta Thunderstorms”. Weather and Forecasting. 17 (5): 1048–1062. Bibcode:2002WtFor..17.1048B. doi:10.1175/1520-0434(2002)017<1048:MMHSIA>2.0.CO;2.
34. Jacque Marshall (2000-04-10). “Hail Fact Sheet”. University Corporation for Atmospheric Research. Retrieved 2009-07-15.
35. M. Klesius (2007). “The Mystery of Snowflakes”. National Geographic. 211 (1): 20. ISSN 0027-9358.
36. William J. Broad (2007-03-20). “Giant Snowflakes as Big as Frisbees? Could Be”. New York Times. Retrieved 2009-07-12.
37. Jennifer E. Lawson (2001). Hands-on Science: Light, Physical Science (matter) - Chapter 5: The Colors of Light. Portage & Main Press. p. 39. ISBN 978-1-894110-63-1. Retrieved 2009-06-28.
38. Kenneth G. Libbrecht (2006-09-11). “Guide to Snowflakes”. California Institute of Technology. Retrieved 2009-06-28.
39. John Roach (2007-02-13). “”No Two Snowflakes the Same” Likely True, Research Reveals”. National Geographic News. Retrieved 2009-07-14.
40. Kenneth Libbrecht (Winter 2004–2005). “Snowflake Science” (PDF). American Educator. Retrieved 2009-07-14.
41. Glossary of Meteorology (June 2000). “Diamond Dust”. American Meteorological Society. Retrieved 2010-01-21.
42. Kenneth G. Libbrecht (2001). “Morphogenesis on Ice: The Physics of Snow Crystals” (PDF). Engineering & Science. California Institute of Technology (1): 12. Retrieved 2010-01-21.
43. B. Geerts (2002). “Convective and stratiform rainfall in the tropics”. University of Wyoming. Retrieved 2007-11-27.
44. David Roth (2006). “Unified Surface Analysis Manual” (PDF). Hydrometeorological Prediction Center. Retrieved 2006-10-22.
45. Glossary of Meteorology (2009). “Graupel”. American Meteorological Society. Retrieved 2009-01-02.

46. Toby N. Carlson (1991). Mid-latitude Weather Systems. Routledge. p. 216. ISBN 978-0-04-551115-0. Retrieved 2009-02-07.
47. Diana Leone (2002). "Rain supreme". Honolulu Star-Bulletin. Retrieved 2008-03-19.
48. Western Regional Climate Center (2002). "Climate of Hawaii". Retrieved 2008-03-19.
49. Paul E. Lydolph (1985). The Climate of the Earth. Rowman & Littlefield. p. 333. ISBN 978-0-86598-119-5. Retrieved 2009-01-02.
50. Michael A. Mares (1999). Encyclopedia of Deserts. University of Oklahoma Press. p. 252. ISBN 978-0-8061-3146-7. Retrieved 2009-01-02.
51. Adam Ganson (2003). "Geology of Death Valley". Indiana University. Retrieved 2009-02-07.
52. Joan Von Ahn; Joe Sienkiewicz; Greggory McFadden (April 2005). "Hurricane Force Extratropical Cyclones Observed Using QuikSCAT Near Real Time Winds". Mariners Weather Log. Voluntary Observing Ship Program. 49 (1). Retrieved 2009-07-07.
53. Owen Hertzman (1988). "Three-Dimensional Kinematics of Rainbands in Midlatitude Cyclones Abstract". PhD thesis. University of Washington. Bibcode:1988PhDT.......110H.
54. Yuh-Lang Lin (2007). Mesoscale Dynamics. Cambridge University Press. p. 405. ISBN 978-0-521-80875-0. Retrieved 2009-07-07.
55. B. Geerts (1998). "Lake Effect Snow.". University of Wyoming. Retrieved 2008-12-24.
56. Greg Byrd (1998-06-03). "Lake Effect Snow". University Corporation for Atmospheric Research. Retrieved 2009-07-12.
57. Karl W. Birkeland & Cary J. Mock (1996). "Atmospheric Circulation Patterns Associated With Heavy Snowfall Events, Bridger Bowl, Montana, USA" (PDF). Mountain Research and Development. International Mountain Society. 16 (3): 281–286. doi:10.2307/3673951. JSTOR 3673951.
58. Glossary of Meteorology (2009). "Rainy season". American Meteorological Society. Retrieved 2008-12-27.
59. Costa Rica Guide (2005). "When to Travel to Costa Rica". ToucanGuides. Retrieved 2008-12-27.
60. Michael Pidwirny (2008). "CHAPTER 9: Introduction to the Biosphere". PhysicalGeography.net. Retrieved 2008-12-27.
61. Elisabeth M. Benders-Hyde (2003). "World Climates". Blue Planet Biomes. Retrieved 2008-12-27.
62. Mei Zheng (2000). "The sources and characteristics of atmospheric particulates during the wet and dry seasons in Hong Kong". University of Rhode Island. Retrieved 2008-12-27.

63. S. I. Efe; F. E. Ogban; M. J. Horsfall; E. E. Akporhonor (2005). "Seasonal Variations of Physico-chemical Characteristics in Water Resources Quality in Western Niger Delta Region, Nigeria" (PDF). Journal of Applied Scientific Environmental Management. 9 (1): 191–195. ISSN 1119-8362. Retrieved 2008-12-27.
64. C. D. Haynes; M. G. Ridpath; M. A. J. Williams (1991). Monsoonal Australia. Taylor & Francis. p. 90. ISBN 978-90-6191-638-3. Retrieved 2008-12-27.
65. Marti J. Van Liere, Eric-Alain D. Ategbo, Jan Hoorweg, Adel P. Den Hartog, and Joseph G. A. J. Hautvast (1994). "The significance of socio-economic characteristics for adult seasonal body-weight fluctuations: a study in north-western Benin" (PDF). British Journal of Nutrition. Cambridge University Press. 72 (3): 479–488. doi:10.1079/BJN19940049. PMID 7947661.
66. Chris Landsea (2007). "Subject: D3 - Why do tropical cyclones' winds rotate counter-clockwise (clockwise) in the Northern (Southern) Hemisphere?". National Hurricane Center. Retrieved 2009-01-02.
67. Climate Prediction Center (2005). "2005 Tropical Eastern North Pacific Hurricane Outlook". National Oceanic and Atmospheric Administration. Retrieved 2006-05-02.
68. Jack Williams (2005-05-17). "Background: California's tropical storms". USA Today. Retrieved 2009-02-07.
69. National Climatic Data Center (2005-08-09). "Global Measured Extremes of Temperature and Precipitation". National Oceanic and Atmospheric Administration. Retrieved 2007-01-18.
70. Dr. Owen E. Thompson (1996). Hadley Circulation Cell. Channel Video Productions. Retrieved on 2007-02-11.
71. ThinkQuest team 26634 (1999). The Formation of Deserts. Oracle ThinkQuest Education Foundation. Retrieved on 2009-02-16.
72. "USGS 220427159300201 1047.0 Mt. Waialeale Rain Gage nr Lihue, Kauai, HI". USGS Real-time rainfall data at Wai□ale□ale Raingauge. Retrieved 2008-12-11.
73. USA Today. Mt. Baker snowfall record sticks. Retrieved on 2008-02-29.
74. National Weather Service Office, Northern Indiana (2009). "8 Inch Non-Recording Standard Rain Gauge". Retrieved 2009-01-02.
75. Chris Lehmann (2009). "10/00". Central Analytical Laboratory. Retrieved 2009-01-02.
76. National Weather Service Office Binghamton, New York (2009). "Rainguage Information". Retrieved 2009-01-02.
77. National Weather Service (2009). "Glossary: W". Retrieved 2009-01-01.
78. Discovery School (2009). "Build Your Own Weather Station". Discovery Education. Archived from the original on 2008-12-26. Retrieved 2009-01-02.
79. "Community Collaborative Rain, Hail & Snow Network Main Page". Colorado Climate Center. 2009. Retrieved 2009-01-02.

80. The Globe Program (2009). "Global Learning and Observations to Benefit the Environment Program". Retrieved 2009-01-02.

81. National Weather Service (2009). "NOAA's National Weather Service Main Page". Retrieved 2009-01-01.

82. Glossary of Meteorology (2009). "Hydrometeor". American Meteorological Society. Retrieved 2009-07-16.

83. National Aeronautics and Space Administration (2012). "NASA and JAXA's GPM Mission Takes Rain Measurements Global". Retrieved 2014-01-21.

84. C. Kidd; G.J. Huffman (2011). "Global Precipitation Measurement". Meteorological Applications. 18 (3): 334–353. Bibcode:2011MeApp..18..334K. doi:10.1002/met.284.

85. F.J. Tapiador, F.J. Turk, W. Petersen, A.Y. Hou, E. Garcia-Ortega, L.T. Machado, C.F. Angelis, P. Salio, C. Kidd, G.J. Hffman M. De Castro (2012). "Global Precipitation Measurement Methods, Datasets and Applications.". Atmospheric Research. 104-105: 70–97. Bibcode:2013AtmRe.119..131W. doi:10.1016/j.atmosres.2011.10.012.

86. International Precipitation Working Group. "Global Precipitation Datasets". Retrieved 2014-01-21.

87. Glossary of Meteorology (June 2000). "Return period". American Meteorological Society. Retrieved 2009-01-02.

88. Glossary of Meteorology (June 2000). "Rainfall intensity return period". American Meteorological Society. Retrieved 2009-01-02.

89. Boulder Area Sustainability Information Network (2005). "What is a 100 year flood?". Boulder Community Network. Retrieved 2009-01-02.

90. Peel, M. C. and Finlayson, B. L. and McMahon, T. A. (2007). "Updated world map of the Köppen-Geiger climate classification". Hydrol. Earth Syst. Sci. 11 (5): 1633–1644. doi:10.5194/hess-11-1633-2007. ISSN 1027-5606. (direct: Final Revised Paper)

91. Susan Woodward (1997-10-29). "Tropical Broadleaf Evergreen Forest: The Rainforest". Radford University. Retrieved 2008-03-14.

92. Susan Woodward (2005-02-02). "Tropical Savannas". Radford University. Retrieved 2008-03-16.

93. "Humid subtropical climate". Encyclopædia Britannica. Encyclopædia Britannica Online. 2008. Retrieved 2008-05-14.

94. Michael Ritter (2008-12-24). "Humid Subtropical Climate". University of Wisconsin–Stevens Point. Retrieved 2008-03-16.

95. Lauren Springer Ogden (2008). Plant-Driven Design. Timber Press. p. 78. ISBN 978-0-88192-877-8.

96. Michael Ritter (2008-12-24). "Mediterranean or Dry Summer Subtropical Climate". University of Wisconsin–Stevens Point. Retrieved 2009-07-17.

97. Brynn Schaffner & Kenneth Robinson (2003-06-06). “Steppe Climate”. West Tisbury Elementary School. Retrieved 2008-04-15.

98. Michael Ritter (2008-12-24). “Subarctic Climate”. University of Wisconsin–Stevens Point. Retrieved 2008-04-16.

99. Bureau of Meteorology (2010). “Living With Drought”. Commonwealth of Australia. Retrieved 2010-01-15.

100. Robert Burns (2007-06-06). “Texas Crop and Weather”. Texas A&M University. Retrieved 2010-01-15.

101. James D. Mauseth (2006-07-07). “Mauseth Research: Cacti”. University of Texas. Retrieved 2010-01-15.

102. A. Roberto Frisancho (1993). Human Adaptation and Accommodation. University of Michigan Press, pp. 388. ISBN 978-0-472-09511-7. Retrieved on 2008-12-27.

103. Climate Change Division (2008-12-17). “Precipitation and Storm Changes”. United States Environmental Protection Agency. Retrieved 2009-07-17.

104. Dale Fuchs (2005-06-28). “Spain goes hi-tech to beat drought”. London: The Guardian. Retrieved 2007-08-02.

105. Goddard Space Flight Center (2002-06-18). “NASA Satellite Confirms Urban Heat Islands Increase Rainfall Around Cities”. National Aeronautics and Space Administration. Archived from the original on March 16, 2010. Retrieved 2009-07-17.

106. Jack S. Bushong (1999). “Quantitative Precipitation Forecast: Its Generation and Verification at the Southeast River Forecast Center”(PDF). University of Georgia. Retrieved 2008-12-31.

107. Daniel Weygand (2008). “Optimizing Output From QPF Helper”(PDF). National Weather Service Western Region. Retrieved 2008-12-31.

108. Noreen O. Schwein (2009). “Optimization of quantitative precipitation forecast time horizons used in river forecasts”. American Meteorological Society. Retrieved 2008-12-31.

109. Christian Keil; Andreas Röpnack; George C. Craig & Ulrich Schumann (2008-12-31). “Sensitivity of quantitative precipitation forecast to height dependent changes in humidity”. Geophysical Research Letters. 35 (9): L09812. Bibcode:2008GeoRL..3509812K. doi:10.1029/2008GL033657.

110. P. Reggiani & A. H. Weerts (2007). “Probabilistic Quantitative Precipitation Forecast for Flood Prediction: An Application”. Journal of Hydrometeorology. 9 (1): 76–95. Bibcode:2008JHyMe...9...76R. doi:10.1175/2007JHM858.1. Retrieved 2008-12-31.

111. Charles Lin (2005). “Quantitative Precipitation Forecast (QPF) from Weather Prediction Models and Radar Nowcasts, and Atmospheric Hydrological Modelling for Flood Simulation” (PDF). Achieving Technological Innovation in Flood Forecasting Project. Retrieved 2009-01-01.

112. “The Monarch With No Gumption Attends the Funeral of the Frisbee”. Retrieved 28 June 2016.

113. Peel, M. C.; Finlayson, B. L.; McMahon, T. A. (2007). "Updated world map of the Köppen–Geiger climate classification". Hydrol. Earth Syst. Sci. 11: 1633–1644. doi:10.5194/hess-11-1633-2007. ISSN 1027-5606. (direct: Final Revised Paper)

114. Köppen, Wladimir (1884). Translated by Volken, E.; Brönnimann, S. "Die Wärmezonen der Erde, nach der Dauer der heissen, gemässigten und kalten Zeit und nach der Wirkung der Wärme auf die organische Welt betrachtet" [The thermal zones of the earth according to the duration of hot, moderate and cold periods and to the impact of heat on the organic world)]. Meteorologische Zeitschrift (published 2011). 20 (3): 351–360 – via http://www.ingentaconnect.com/content/schweiz/mz/2011/00000020/00000003/art00009.

115. Rubel, F.; Kottek, M (2011). "Comments on: 'The thermal zones of the Earth' by Wladimir Köppen (1884)". Meteorologische Zeitschrift. 20 (3): 361–365.

116. Köppen, Wladimir (1918). "Klassification der Klimate nach Temperatur, Niederschlag and Jahreslauf". Petermanns Geographische Mitteilungen. 64. pp. 193–203, 243–248 – via http://koeppen-geiger.vu-wien.ac.at/koeppen.htm.

117. Köppen, Wladimir (1936). "C". In Köppen, Wladimir; Geiger (publisher), Rudolf. Das geographische System der Klimate [The geographic system of climates] (PDF). Handbuch der Klimatologie. 1. Berlin: Borntraeger.

118. Geiger, Rudolf (1954). "Klassifikation der Klimate nach W. Köppen" [Classification of climates after W. Köppen]. Landolt-Börnstein – Zahlenwerte und Funktionen aus Physik, Chemie, Astronomie, Geophysik und Technik, alte Serie. Berlin: Springer. 3. pp. 603–607.

119. Geiger, Rudolf (1961). Überarbeitete Neuausgabe von Geiger, R.: Köppen-Geiger / Klima der Erde. (Wandkarte 1:16 Mill.) – Klett-Perthes, Gotha.

120. McKnight, Tom L; Hess, Darrel (2000). "Climate Zones and Types". Physical Geography: A Landscape Appreciation. Upper Saddle River, NJ: Prentice Hall. ISBN 0-13-020263-0.

121. Linacre, Edward; Bart Geerts (1997). Climates and Weather Explained. London: Routledge. p. 379. ISBN 0-415-12519-7.

122. Critchfield, H.J. (1983). "Criteria for classification of major climatic types in modified Köppen system" (4 ed.). University of Idaho.

123. "Mediterranean Climate". California Rangelands. California Ragelands. Retrieved 2015-01-26.

124. http://www.climate.weatheroffice.gc.ca/climate_normals/results_e.html?stnID=823&lang=e&dCode=0&province=BC&provBut=Search&month1=0&month2=12

125. "Iceland Met office: Monthly Averages for Reykjavík". Iceland Met Office. 2012. Retrieved January 4, 2013.

126. http://images.slideplayer.es/1/101786/slides/slide_3.jpg

127. Chen, D.; Chen, H. W. (2013). "Using the Köppen classification to quantify climate variation and change: An example for 1901–2010". Environmental Development. 6: 69–79. doi:10.1016/j.envdev.2013.03.007. (direct: Final Revised Paper)

128. http://hanschen.org/koppen/
129. Critchfield, Howard J (1983). General Climatology (4th ed.). New Delhi: Prentice Hall. pp. 154–161. ISBN 978-81-203-0476-5.
130. http://www.nature.com/articles/srep13487
131. "Quantification of cloud water interception". Elsevier. Retrieved 2008-08-18.
132. "METAR Conversion Card". National Weather Service. Retrieved 2012-12-12.
133. B. Geerts. Convective and stratiform rainfall in the tropics. Retrieved on 2007-11-27.
134. Houze, Robert (October 1997). "Stratiform Precipitation in Regions of Convection: A Meteorological Paradox?" (PDF). Bulletin of the American Meteorological Society. 78 (10): 2179. Bibcode:1997BAMS...78.2179H. doi:10.1175/1520-0477(1997)078<2179:SPIROC>2.0.CO;2. ISSN 1520-0477. Retrieved 2007-11-27.[dead link]
135. Glossary of Meteorology. Graupel. Retrieved on 2009-01-02.
136. Glossary of Meteorology (2009). "Stratiform precipitation area". American Meteorological Society. Retrieved 2009-07-12.
137. Physical Geography. CHAPTER 8: Introduction to the Hydrosphere (e). Cloud Formation Processes. Retrieved on 2009-01-01.
138. Diana Leone Rain supreme. Retrieved on 2008-03-19.
139. Western Regional Climate Center. Climate of Hawaii. Retrieved on 2008-03-19.
140. Paul E. Lydolph. The Climate of the Earth. Retrieved on 2009-01-02.
141. Michael A. Mares. Encyclopedia of Deserts. Retrieved on 2009-01-02.
142. Glossary of Meteorology (2012). "Rain". American Meteorological Society. Retrieved 2014-11-13.
143. Glossary of Meteorology (2009). "Snow". American Meteorological Society. Retrieved 2009-06-28.
144. Samuel Miller. "Lesson 7: Clouds and Precipitation". Retrieved 2011-07-08.
145. David Roth. "Unified Surface Analysis Manual" (PDF). Hydrometeorological Prediction Center. Retrieved 2006-10-22.
146. Glossary of Meteorology. Airmass Classification. Retrieved on 2008-05-22.
147. C. Donald Ahrens (2007). Meteorology today: an introduction to weather, climate, and the environment. Cengage Learning. p. 296. ISBN 978-0-495-01162-0. Retrieved 2010-12-17.
148. Monmonier, Mark. Air Apparent: How Meteorologists Learned to Map, Predict, and Dramatize Weather. University of Chicago Press. Chicago: 1999.
149. Chris C. Park (2001). The environment: principles and applications. Psychology Press. p. 309. ISBN 978-0-415-21771-2. Retrieved 2010-12-17.
150. "Overrunning". NWS Glossary. National Weather Service. Retrieved 2010-05-02.
151. "Occluded Front". University of Illinois Department of Atmospheric Sciences. Retrieved 2006-10-22.

152. National Weather Service Office, Norman, Oklahoma. "Triple Point". NOAA. Retrieved 2006-10-22.
153. "Trowal". World Meteorological Organisation. Eumetcal. Retrieved 2013-08-28.
154. Stationary Front. University of Illinois Department of Atmospheric Sciences. Retrieved on 2006-10-22.
155. "Shear Line". Glossary of Meteorology. American Meteorological Society. Retrieved 2006-10-22.
156. Huaqing Cai. "Dryline cross section". Retrieved 2006-12-05.
157. "Lee Trough". Glossary of Meteorology. American Meteorological Society. Retrieved 2006-10-22.
158. "Dry Line: A Moisture Boundary". University of Illinois Department of Atmospheric Science. Retrieved 2006-10-22.
159. Office of the Federal Coordinator for Meteorology. "Chapter 2: Definitions" (PDF). Retrieved 2006-10-22.
160. Michael Branick. "A Comprehensive Glossary of Weather". American Meteorological Society. Retrieved 2006-10-22.
161. Western Region Climate Center. "H". Retrieved 2006-10-22.
162. "Convection". Glossary of Meteorology. American Meteorological Society. Retrieved 2006-10-22.
163. "Orographic Lifting". Glossary of Meteorology. American Meteorological Society. Retrieved 2006-10-22.

CHAPTER 5

Infiltration Of Hydrology

Chapter Contents

- **Introduction**
- **Process**
- **Infiltration Calculation Methods**

Infiltration is the process by which precipitation or water soaks into subsurface soils and moves into rocks through cracks and pore spaces. As we mentioned before, the bulk of rainwater and melted snow end up infiltrated.

INTRODUCTION

Infiltration is the process by which water on the ground surface enters the soil. Infiltration rate in soil science is a measure of the rate at which soil is able to absorb rainfall or irrigation. It is measured in inches per hour or millimeters per hour. The rate decreases as the soil becomes saturated. If the precipitation rate exceeds the infiltration rate, runoff will usually occur unless there is some physical barrier. It is related to the saturated hydraulic conductivity of the near-surface soil. The rate of infiltration can be measured using an infiltrometer.

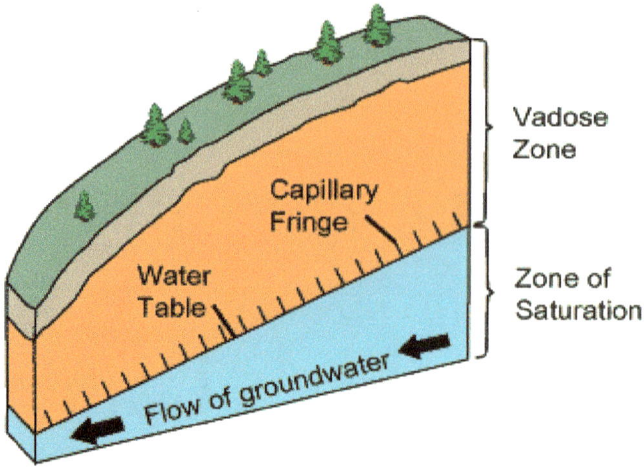

Cross-section of a hillslope depicting the vadose zone, capillary fringe, water table, and phreatic or saturated zone. (Source: United States Geological Survey.)

Infiltration is caused by two forces: gravity and capillary action. While smaller pores offer greater resistance to gravity, very small pores pull water through capillary action in addition to and even against the force of gravity.

The rate of infiltration is determined by soil characteristics including ease of entry, storage capacity, and transmission rate through the soil. The soil texture and structure, vegetation types and cover, water content of the soil, soil temperature, and rainfall intensity all play a role in controlling infiltration rate and capacity. For example, coarse-grained sandy soils have large spaces between each grain and allow water to infiltrate quickly. Vegetation creates more porous soils by protecting the soil from raindrop impact, which can close natural gaps between soil particles, loosening soil through root action and enhancing the presence of soil organism like termites, worms and small mammals that have a direct impact on soil bulk densities. This is why forested areas have the highest infiltration rates of any vegetative types.

The top layer of leaf litter that is not decomposed protects the soil from the pounding action of rain; without this the soil can become far less permeable. In chaparral vegetated areas, the hydrophobic oils in the succulent leaves can be spread over the soil surface with fire, creating large areas of hydrophobic soil. Other conditions that can lower infiltration rates or block them include dry plant litter that resists re-wetting, or frost. If soil is saturated at the time of an intense freezing period, the soil can become a concrete frost on which almost no infiltration would occur. Over an entire watershed, there are likely to be gaps in the concrete frost or hygroscopic soil where water could infiltrate. Once water has infiltrated the soil it remains in the soil, percolates down to the ground water table, or becomes part of the subsurface runoff process.

PROCESS

The process of infiltration can continue only if there is room available for additional water at the soil surface. The available volume for additional water in the soil depends on the porosity of the soil and the rate at which previously infiltrated water can move away from the surface through the soil. The maximum rate that water can enter a soil in a given condition is the infiltration capacity. If the arrival of the water at the soil surface is less than the infiltration capacity, it is sometimes analyzed using hydrology transport models, mathematical models that consider infiltration, runoff and channel flow to predict river flow rates and stream water quality.

Research Findings

Robert E. Horton suggested that infiltration capacity rapidly declines during the early part of a storm and then tends towards an approximately constant value after a couple of hours for the remainder of the event. Previously infiltrated water fills the available storage spaces and reduces the capillary forces drawing water into the pores. Clay particles in the soil may swell as they become wet and thereby reduce the size of the pores. In areas where the ground is not protected by a layer of forest litter, raindrops can detach soil particles from the surface and wash fine particles into surface pores where they can impede the infiltration process.

Infiltration in Wastewater Collection

Wastewater collection systems consist of a set of lines, junctions and lift stations to convey sewage to a wastewater treatment plant. When these Herr lines are compromised by rupture, cracking or tree root invasion, infiltration/inflow of stormwater often occurs. This circumstance can lead to a sanitary sewer overflow, or discharge of untreated sewage to the environment.

INFILTRATION CALCULATION METHODS

Infiltration is a component of the general mass balance hydrologic budget. There are several ways to estimate the volume and/or the rate of infiltration of water into a soil. The rigorous standard that fully couples groundwater to surface water through a non-homogeneous soil is the numerical solution of Richards' equation. A newer method that allows full groundwater and surface water coupling in homogeneous soil layers, and that is related to the Richards equation is the Finite water-content vadose zone flow method. In the case of uniform initial soil water content and a deep well-drained soil, there are some excellent approximate methods to solve for the infiltration flux for a single rainfall event. Among these are the Green and Ampt (1911) method, Parlange et al. (1982). Beyond these methods there are a host of empirical methods such as, SCS method, Horton's method, etc., that are little more than curve fitting exercises.

General Hydrologic Budget

The general hydrologic budget, with all the components, with respect to infiltration *F*. Given all the other variables and infiltration is the only unknown, simple algebra solves the infiltration question.

$$F = B_I + P - E - T - ET - S - I_A - R - B_O$$

where

F is infiltration, which can be measured as a volume or length;

B_I is the boundary input, which is essentially the output watershed from adjacent, directly connected impervious areas;

B_O is the boundary output, which is also related to surface runoff, R, depending on where one chooses to define the exit point or points for the boundary output;

P is precipitation;

E is evaporation;

T is transpiration;

ET is evapotranspiration;

S is the storage through either retention or detention areas;

I_A is the initial abstraction, which is the short term surface storage such as puddles or even possibly detention ponds depending on size;

R is surface runoff.

The only note on this method is one must be wise about which variables to use and which to omit, for doubles can easily be encountered. An easy example of double counting variables is when the evaporation, E, and the transpiration, T, are placed in the equation as well as the evapotranspiration, ET. ET has included in it T as well as a portion of E. Interception also needs to be accounted for, not just raw precipitation.

Richards' Equation (1931)

The standard rigorous approach for calculating infiltration into soils is Richards' Equation, which is a partial differential equation with very nonlinear coefficients. The Richards equation is computationally expensive, not guaranteed to converge, and sometimes has difficulty with mass conservation.

Finite Water-Content Vadose Zone Flow Method

This method is an approximation of the Richards' (1931) partial differential equation that de-emphasized soil water diffusivity and emphasizes advection. This approximation does not affect the calculated infiltration flux because the diffusive flux has a mean of 0. The Finite water-content vadose zone flow method is a set of three ordinary differential equations, is guaranteed to converge and to conserve mass. It requires the assumption that soil is uniform within layers.

Green and Ampt

Named for two men; Green and Ampt. The Green-Ampt method of infiltration estimation accounts for many variables that other methods, such as Darcy's law, do not. It is a function of the soil suction head, porosity, hydraulic conductivity and time.

$$\int_0^{F(t)} \frac{F}{F+\psi\,\Delta\theta}\,dF = \int_0^t K\,dt$$

where

ψ is wetting front soil suction head (L);

θ is water content (-);

K is Hydraulic conductivity (L/T);

F(t) is the cumulative depth of infiltration (L).

Once integrated, one can easily choose to solve for either volume of infiltration or instantaneous infiltration rate:

$$F(t) = Kt + \psi\,\Delta\theta \ln\left[1 + \frac{F(t)}{\psi\,\Delta\theta}\right].$$

Using this model one can find the volume easily by solving for $F(t)$. However the variable being solved for is in the equation itself so when solving for this one must set the variable in question to converge on zero, or another appropriate constant. A good first guess for F is the larger value of either Kt or $\sqrt{2\psi\,\Delta\theta Kt}$. The only note on using this formula is that one must assume that h_0, the water head or the depth of ponded water above the surface, is negligible. Using the infiltration volume from this equation one may then substitute F into the corresponding infiltration rate equation below to find the instantaneous infiltration rate at the time, t, F was measured.

$$f(t) = K\left[\frac{\psi\,\Delta\theta}{F(t)} + 1\right].$$

Horton's Equation

Named after the same Robert E. Horton mentioned above, Horton's equation is another viable option when measuring ground infiltration rates or volumes. It is an empirical formula that says that infiltration starts at a constant rate, f_0 and is decreasing exponentially with time t After some time when the soil saturation level reaches a certain value, the rate of infiltration will level off to the rate f^c

$$f_t = f_c + (f_0 - f_c)e^{-kt}$$

Where

f_t is the infiltration rate at time t;

f_0 is the initial infiltration rate or maximum infiltration rate;

f_c is the constant or equilibrium infiltration rate after the soil has been saturated or minimum infiltration rate;

k is the decay constant specific to the soil

The other method of using Horton's equation is as below. It can be used to find the total volume of infiltration, F, after time t.

$$F_t = f_c t + \frac{(f_0 - f_c)}{k}(1 - e^{-kt})$$

Kostiakov Equation

Named after its founder Kostiakov is an empirical equation which assumes that the intake rate declines over time according to a power function.

$$f(t) = akt^{a-1}$$

Where a and k are empirical parameters.

The major limitation of this expression is its reliance on the zero final intake rate. In most cases the infiltration rate instead approaches a finite steady value, which in some cases may occur after short periods of time. The Kostiakov-Lewis variant, also known as the "Modified Kostiakov" equation corrects for this by adding a steady intake term to the original equation.

$$f(t) = akt^{a-1} + f_0$$

in integrated form the cumulative volume is expressed as:

$$F(t) = kt^a + f_0 t$$

Where

f_0 approximates, but does not necessarily equate to the final infiltration rate of the soil.

Darcy's Law

This method used for infiltration is using a simplified version of Darcy's law. Many would argue that this method is too simple and should not be used. Compare it with the Green and Ampt (1911) solution mentioned previously. This method is similar to Green and Ampt, but missing the cumulative infiltration depth and is therefore incomplete because it assumes that the infiltration gradient occurs over some arbitrary length L In this model the ponded water is assumed to be equal to h_0 and the head of dry soil that exists below the depth of the wetting front soil suction head is assumed to be equal to $-\psi - L$.

$$f = K\left[\frac{h_0 - (-\psi - L)}{L}\right]$$

Where

ψ is wetting front soil suction head

h_0 is the depth of ponded water above the ground surface;

K is the hydraulic conductivity;

L is the vague total depth of subsurface ground in question. This vague definition explains why this method should be avoided.

Or

$$f = K\left[\frac{L + S_f + h_0}{L}\right]$$

f Infiltration rate f (mm hour−1))

K is the hydraulic conductivity (mm hour−1));

L is the vague total depth of subsurface ground in question (mm). This vague definition explains why this method should be avoided.

S_f is wetting front soil suction head $(-\psi) = (-\psi_f)$ (mm)

h_0 is the depth of ponded water above the ground surface (mm);

REFERENCES

1. Hogan, C. Michael (2010). "Abiotic factor" in Encyclopedia of Earth. eds Emily Monosson and C. Cleveland. National Council for Science and the Environment. Washington DC
2. Horton, Robert E. (1933). "The role of infiltration in the hydrologic cycle". Trans. Am. Geophys. Union. 14th Ann. Mtg: 446–460. Bibcode:1933TrAGU..14..446H. doi:10.1029/TR014i001p00446.
3. Green, W. Heber; Ampt, G. A. (1911). "Studies on Soil Physics". The Journal of Agricultural Science. 4: 1. doi:10.1017/S0021859600001441.
4. Parlange, J. -Y.; Lisle, I.; Braddock, R. D.; Smith, R. E. (1982). "The Three-Parameter Infiltration Equation". Soil Science. 133 (6): 337. doi:10.1097/00010694-198206000-00001.
5. Richards, L. A. (1931). "Capillary Conduction of Liquids Through Porous Mediums". Physics. 1 (5): 318. Bibcode:1931Physi...1..318R. doi:10.1063/1.1745010.
6. Ogden, F. L.; Lai, W.; Steinke, R.C.; Zhu, J.; Talbot, C.A.; Wilson, J.L. (2015). "A new general 1-D vadose zone flow solution method". Water Resour. Res. 51: 4282. Bibcode:2015WRR....51.4282O. doi:10.1002/2015WR017126.
7. Water Resources Engineering, 2005 Edition, John Wiley & Sons, Inc.
8. Kostiakov, A.N. "On the dynamics of the coefficient of water-percolation in soils and on the necessity of studying it from a dynamic point of view for purposes of amelioration". Transactions of 6th Congress of International Soil Science Society. Moscow. pp. 17–21.
9. Walker, W.R.; Skogerboe, G.V. (1987). Surface irrigation: Theory and practice. Prentice-Hall, Englewood Cliffs.
10. Hendriks, Martin R. (2010) Introduction to Physical Hydrology, Oxford University Press

CHAPTER 6

Water Quality

Chapter Contents

- **Introduction**
- **Standards**
- **Categories**
- **Sampling And Measurement**
- **Wastewater Quality Indicators**
- **Water Treatment**

Water quality describes the condition of the water, including chemical, physical, and biological characteristics, usually with respect to its suitability for a particular purpose such as drinking or swimming

Water quality is measured by several factors, such as the concentration of dissolved oxygen, bacteria levels, the amount of salt (or salinity), or the amount of material suspended in the water (turbidity). In some bodies of water, the concentration of microscopic algae and quantities of pesticides, herbicides, heavy metals, and other contaminants may also be measured to determine water quality.

INTRODUCTION

Water quality refers to the chemical, physical, biological, and radiological characteristics of water. It is a measure of the condition of water relative to the requirements of one or more biotic species and or to any human need or purpose. It is most frequently used by reference to a set of standards against which compliance can be assessed. The most common standards used to assess water quality relate to health of ecosystems, safety of human contact, and drinking water.

A rosette sampler is used to collect water samples in deep water, such as the Great Lakes or oceans, for water quality testing.

STANDARDS

In the setting of standards, agencies make political and technical/scientific decisions about how the water will be used. In the case of natural water bodies, they also make some reasonable estimate of pristine conditions. Natural water bodies will vary in response to environmental conditions. Environmental scientists work to understand how these systems function, which in turn helps to identify the sources and fates of contaminants. Environmental lawyers and policymakers work to define legislation with the intention that water is maintained at an appropriate quality for its identified use.

The vast majority of surface water on the Earth is neither potable nor toxic. This remains true when seawater in the oceans (which is too salty to drink) is not counted. Another general perception of water quality is that of a simple property that tells whether water is polluted or not. In fact, water quality is a complex subject, in part because water is a complex medium intrinsically tied to the ecology of the Earth. Industrial and commercial activities (e.g. manufacturing, mining, construction, transport) are a major cause of water pollution as are runoff from agricultural areas, urban runoff and discharge of treated and untreated sewage.

CATEGORIES

The parameters for water quality are determined by the intended use. Work in the area of water quality tends to be focused on water that is treated for human consumption, industrial use, or in the environment.

Human Consumption

Contaminants that may be in untreated water include microorganisms such as viruses, protozoa and bacteria; inorganic contaminants such as salts and metals; organic chemical contaminants from industrial processes and petroleum use; pesticides and herbicides; and radioactive contaminants. Water quality depends on the local geology and ecosystem, as well as human uses such as sewage dispersion, industrial pollution, use of water bodies as a heat sink, and overuse (which may lower the level of the water).

The United States Environmental Protection Agency (EPA) limits the amounts of certain contaminants in tap water provided by US public water systems. The Safe Drinking Water Act authorizes EPA to issue two types of standards:

- *primary standards* regulate substances that potentially affect human health;
- *secondary standards* prescribe aesthetic qualities, those that affect taste, odor, or appearance.

The U.S. Food and Drug Administration (FDA) regulations establish limits for contaminants in bottled water that must provide the same protection for public health. Drinking water, including bottled water, may reasonably be expected to contain at least small amounts of some contaminants. The presence of these contaminants does not necessarily indicate that the water poses a health risk.

In urbanized areas around the world, water purification technology is used in municipal water systems to remove contaminants from the source water (surface water or groundwater) before it is distributed to homes, businesses, schools and other recipients. Water drawn directly from a stream, lake, or aquifer and that has no treatment will be of uncertain quality.

Industrial and Domestic Use

Dissolved minerals may affect suitability of water for a range of industrial and domestic purposes. The most familiar of these is probably the presence of ions of calcium (Ca^{2+}) and magnesium (Mg^{2+}) which interfere with the cleaning action of soap, and can form hard sulfate and soft carbonate deposits in water heaters or boilers. Hard water may be softened to remove these ions. The softening process often substitutes sodium cations. Hard water may be preferable to soft water for human consumption, since health problems have been associated with excess sodium and with calcium and magnesium deficiencies. Softening decreases nutrition and may increase cleaning effectiveness. Various industries' wastes and effluents can also pollute the water quality in receiving bodies of water.

Environmental Water Quality

Environmental water quality, also called ambient water quality, relates to water bodies such as lakes, rivers, and oceans. Water quality standards for surface waters vary significantly due to different environmental conditions, ecosystems, and intended human uses.

Urban runoff discharging to coastal waters

Toxic substances and high populations of certain microorganisms can present a health hazard for non-drinking purposes such as irrigation, swimming, fishing, rafting, boating, and industrial uses. These conditions may also affect wildlife, which use the water for drinking or as a habitat. Modern water quality laws generally specify protection of fisheries and recreational use and require, as a minimum, retention of current quality standards.

Satirical cartoon by William Heath, showing a woman observing monsters in a drop of London water (at the time of the *Commission on the London Water Supply* report, 1828)

There is some desire among the public to return water bodies to pristine, or pre-industrial conditions. Most current environmental laws focus on the designation of particular uses of a water body. In some countries these designations allow for some water contamination as long as the particular type of contamination is not harmful to the designated uses. Given the landscape changes (e.g., land development, urbanization, clearcutting in forested areas) in the watersheds of many freshwater bodies, returning to pristine conditions would be a significant challenge. In these cases, environmental scientists focus on achieving goals for maintaining healthy ecosystems and may concentrate on the protection of populations of endangered species and protecting human health.

SAMPLING AND MEASUREMENT

The complexity of water quality as a subject is reflected in the many types of measurements of water quality indicators. The most accurate measurements of water quality are made on-site, because water exists in equilibrium with its surroundings. Measurements commonly made on-site and in direct contact with the water source in question include temperature, pH, dissolved oxygen, conductivity, oxygen reduction potential (ORP), turbidity, and Secchi disk depth.

Sample Collection

More complex measurements are often made in a laboratory requiring a water sample to be collected, preserved, transported, and analyzed at another location. The process of water sampling introduces two significant problems:

- The first problem is the extent to which the sample may be representative of the water source of interest. Many water sources vary with time and with location. The measurement of interest may vary seasonally or from day to night or in response to some activity of man or natural populations of aquatic plants and animals. The measurement of interest may vary with distances from the water boundary with overlying atmosphere and underlying or confining soil. The sampler must determine if a single time and location meets the needs of the investigation, or if the water use of interest can be satisfactorily assessed by averaged values with time and/or location, or if critical maxima and minima require individual measurements over a range of times, locations and/or events. The sample collection procedure must assure correct weighting of individual sampling times and locations where averaging is appropriate.[:39–40] Where critical maximum or minimum values exist, statistical methods must be applied to observed variation to determine an adequate number of samples to assess probability of exceeding those critical values.

An automated sampling station installed along the East Branch Milwaukee River, New Fane, Wisconsin. The cover of the 24-bottle autosampler (center) is partially raised, showing the sample bottles inside. The autosampler was programmed to collect samples at time intervals, or proportionate to flow over a specified period. The data logger (white cabinet) recorded temperature, specific conductance, and dissolved oxygen levels.

- The second problem occurs as the sample is removed from the water source and begins to establish chemical equilibrium with its new surroundings - the sample container. Sample containers must be made of materials with minimal reactivity with substances to be measured; and pre-cleaning of sample containers is important. The water sample may dissolve part of the sample container and any residue on that container, or chemicals dissolved in the water sample may sorb onto the sample container and remain there when the water is poured out for analysis.[4] Similar physical and chemical interactions may take place with any pumps, piping, or intermediate devices used to transfer the water sample into the sample container. Water collected from depths below the surface will normally be held at the reduced pressure of the atmosphere; so gas dissolved in the water may escape into unfilled space at the top of the container. Atmospheric gas present in that air space may also dissolve into the water sample. Other chemical reaction equilibria may change if the water sample changes temperature. Finely divided solid particles formerly suspended by water turbulence may settle to the bottom of the sample container, or a solid phase may form from biological growth or chemical precipitation. Microorganisms within the water sample may biochemically alter concentrations of oxygen, carbon dioxide, and organic compounds. Changing carbon dioxide concentrations may alter pH and change solubility of chemicals of interest. These problems are of special concern during measurement of chemicals assumed to be significant at very low concentrations.

Filtering a manually collected water sample (grab sample) for analysis

Sample preservation may partially resolve the second problem. A common procedure is keeping samples cold to slow the rate of chemical reactions and phase change, and analyzing the sample as soon as possible; but this merely minimizes the changes rather than preventing them. A useful procedure for determining influence of sample containers during delay between sample collection and analysis involves preparation for two artificial samples in advance of the sampling event. One sample container is filled with water known from previous analysis to contain no detectable amount of the chemical of interest. This sample, called a "blank", is opened for exposure to the atmosphere when the sample of interest is collected, then resealed and transported to the laboratory with the sample for analysis to determine if sample holding procedures introduced any measurable amount of the chemical of interest. The second artificial sample is collected with the sample of interest, but then "spiked" with a measured additional amount of the chemical of interest at the time of collection. The blank and spiked samples are carried with the sample of interest and analyzed by the same methods at the same times to determine any changes indicating gains or losses during the elapsed time between collection and analysis.

Testing in Response to Natural Disasters and Other Emergencies

Inevitably after events such as earthquakes and tsunamis, there is an immediate response by the aid agencies as relief operations get underway to try and restore basic infrastructure and provide the basic fundamental items that are necessary for survival and subsequent recovery. Access to clean drinking water and adequate sanitation is a priority at times like this. The threat of disease increases hugely due to the large numbers of people living close together, often in squalid conditions, and without proper sanitation.

After a natural disaster, as far as water quality testing is concerned there are widespread views on the best course of action to take and a variety of methods can be employed. The key basic water quality parameters that need to be addressed in an emergency are bacteriological indicators of fecal contamination, free chlorine residual, pH, turbidity and possibly conductivity/total dissolved solids. There are a number of portable water test kits on the market widely used by aid and relief agencies for carrying out such testing.

After major natural disasters, a considerable length of time might pass before water quality returns to pre-disaster levels. For example, following the 2004 Indian Ocean Tsunami the Colombo-based International Water Management Institute (IWMI) monitored the effects of saltwater and concluded that the wells recovered to pre-tsunami drinking water quality one and a half years after the event. IWMI developed protocols for cleaning wells contaminated by saltwater; these were subsequently officially endorsed by the World Health Organization as part of its series of Emergency Guidelines.

Chemical Analysis

The simplest methods of chemical analysis are those measuring chemical elements without respect to their form. Elemental analysis for oxygen, as an example, would indicate a concentration of 890,000 milligrams per litre (mg/L) of water sample because water is made of oxygen. The method selected to measure dissolved oxygen should differentiate between diatomic oxygen and oxygen combined with other elements.

A gas chromatograph-mass spectrometer measures pesticides and other organic pollutants

The comparative simplicity of elemental analysis has produced a large amount of sample data and water quality criteria for elements sometimes identified as heavy metals. Water analysis for heavy metals must consider soil particles suspended in the water sample. These suspended soil particles may contain measurable amounts of metal. Although the particles are not dissolved in the water, they may be consumed by people drinking the water. Adding acid to a water sample to prevent loss of dissolved metals onto the sample container may dissolve more metals from suspended soil particles. Filtration of soil particles from the water sample before acid addition, however, may cause loss of dissolved metals onto the filter. The complexities of differentiating similar organic molecules are even more challenging.

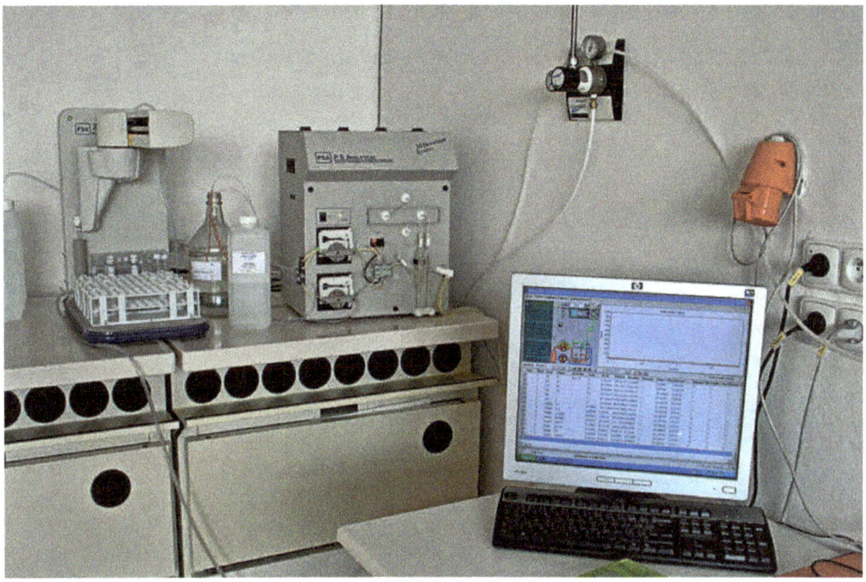

Atomic fluorescence spectroscopy is used to measure mercury and other heavy metals

Making these complex measurements can be expensive. Because direct measurements of water quality can be expensive, ongoing monitoring programs are typically conducted by government agencies. However, there are local volunteer programs and resources available for some general assessment. Tools available to the general public include on-site test kits, commonly used for home fish tanks, and biological assessment procedures.

Real-Time Monitoring

Although water quality is usually sampled and analyzed at laboratories, nowadays, citizens demand real-time information about the water they are drinking. During the last years, several companies are deploying worldwide real-time remote monitoring systems for measuring water pH, turbidity or dissolved oxygen levels.

Drinking Water Indicators

The following is a list of indicators often measured by situational category:

- Alkalinity
- Color of water
- pH

- Taste and odor (geosmin, 2-Methylisoborneol (MIB), etc.)
- Dissolved metals and salts (sodium, chloride, potassium, calcium, manganese, magnesium)
- Microorganisms such as fecal coliform bacteria (*Escherichia coli*), Cryptosporidium, and Giardia lamblia;
- Dissolved metals and metalloids (lead, mercury, arsenic, etc.)
- Dissolved organics: colored dissolved organic matter (CDOM), dissolved organic carbon (DOC)
- Radon
- Heavy metals
- Pharmaceuticals
- Hormone analogs

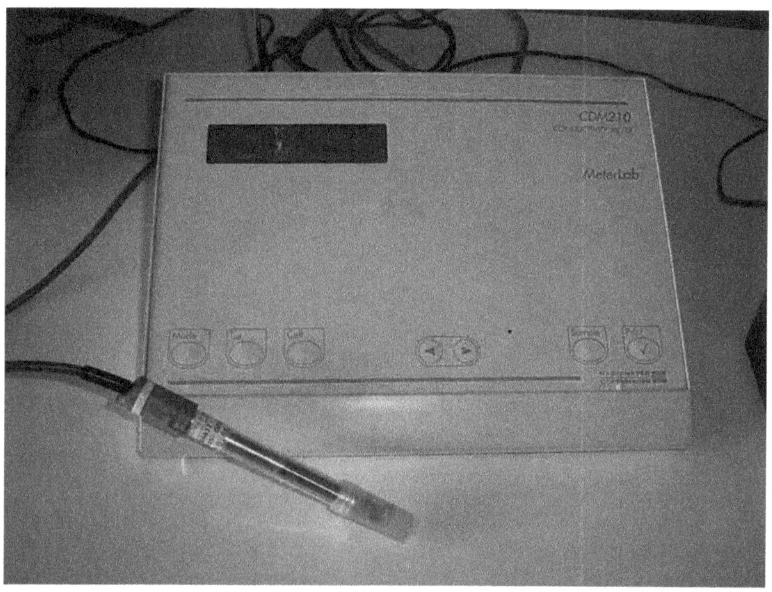

An electrical conductivity meter is used to measure total dissolved solids

Environmental Indicators

Environmental indicators are simple measures that tell us what is happening in the environment. Since the environment is very complex, indicators provide a more practical and economical way to track the state of the environment than if we attempted to record every possible variable in the environment. For example, concentrations of ozone depleting substances (ODS) in the atmosphere, tracked over time, is a good indicator with respect to the environmental issue of stratospheric ozone depletion..

Environmental indicators have been defined in different ways but common themes exist.

"An environmental indicator is a numerical value that helps provide insight into the state of the environment or human health. Indicators are developed based on quantitative measurements or statistics of environmental condition that are tracked over time. Environmental indicators can be developed and used at a wide variety of geographic scales, from local to regional to national levels."

"A parameter or a value derived from parameters that describe the state of the environment and its impact on human beings, ecosystems and materials, the pressures on the environment, the driving forces and the responses steering that system. An indicator has gone through a selection and/or aggregation process to enable it to steer action."

Discussion

Environmental indicator criteria and frameworks have been used to help in their selection and presentation.

It can be considered, for example, that there are major subsets of environmental indicators in-line with the Pressure-State-Response model developed by the OECD. One subset of environmental indicators is the collection of ecological indicators which can include physical, biological and chemical measures such as atmospheric temperature, the concentration of ozone in the stratosphere or the number of breeding bird pairs in an area. These are also referred to as "state" indicators as their focus is on the state of the environment or conditions in the environment. A second subset is the collection of indicators that measure human activities or anthropogenic pressures, such as greenhouse gas emissions. These are also referred to as "pressure" indicators. Finally, there are indicators, such as the number of people serviced by sewage treatment, which track societal responses to environmental issues.

Environmental indicators, in turn, should be considered as a subset of sustainable development indicators which are meant to track the overall sustainability of a society with respect to its environmental, social and economic integrity and health.

A common framework spearheaded by the European Environment Agency is the "DPSIR" or "drivers, pressures, state, impact, response" framework. Drivers and pressures are indicators of the human activities and resulting pressures on the environment in the form of pollution or land-use change, for example. State and impact indicators are the resulting conditions in the environment and the implications for the health of ecosystems and humans. The response indicators measure the reaction of human society to the environmental issue. Criteria tend to focus on three key areas – scientific credibility, policy/social relevance and practical monitoring and data requirements.

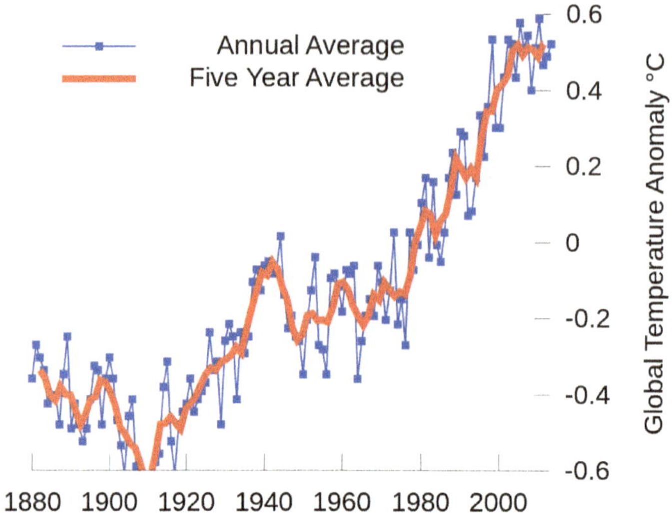

An example of an environmental indicator: Trend in global temperature anomalies of the last 150 years as an indicator of climate change

Environmental indicators are used by governments, non-government organizations, community groups and research institutions to see if environmental objectives are being met, to communicate the state of the environment to the general public and decision makers and as a diagnostic tool through detecting trends in the environment.

Environmental indicators can be measured and reported at different scales. For example, a town may track air quality along with water quality and count the number of rare species of birds to estimate the health of the environment in their area. Indicators are developed for specific ecosystems, such as the Great-Lakes in North America. National governments use environmental indicators to show status and trends with respect to environmental issues of importance to their citizens.

Use in Assessment

Some have attempted to monitor and assess the state of the planet using indicators.

Lester Brown of the Earth Policy Institute, has said: —

Environmentally, the world is in an overshoot mode. If we use environmental indicators to evaluate our situation, then the global decline of the economy's natural support systems - the environmental decline that will lead to economic decline and social collapse- is well underway.

Environmental indicators are also used by companies in the framework of an Environmental management system. The EU Eco-Management and Audit Scheme provides core indicators or Performance Indicator (KPIs) with which registered organizations can measure their environmental performance and monitor their continual environmental improvement against set targets.

Audience

The types of indicators selected or developed should be partially based on who will be using the information from the indicators. There are generally three possible audiences to consider, each with different information needs. These audiences are: 1) technical experts and science advisors, 2) policy-makers, decision makers and resource managers, and 3) general public and media.

The technical experts and scientists will be interested in detailed and complex indicators. These indicators should have scientific validity, sensitivity, responsiveness and have data available on past conditions. The audience that includes policy-makers and resource managers will be concerned with using indicators that are directly related to evaluating policies and objectives. They require their indicators to be sensitive, responsive and have historical data available like the technical audience, but they are also looking for indicators that are cost-effective and have meaning for public awareness. Finally, the general public responds to indicators that have clear and simple messages and are meaningful to them, such as the UV index and the air quality index.

Indicator systems and communicating them

Individual indicators are designed to translate complex information in a concise and easily understood manner in order to represent a particular phenomenon (e.g. ambient air quality). In contrast, indicator systems (or collections of indicators), when seen as a whole are meant to provide an assessment of the full environment domain or a major subset of it (e.g. forests).

Some indicator systems have evolved to include many indicators and require a certain level of knowledge and expertise in various disciplines to fully grasp. A number of methods have been devised in the recent past to boil down this information and allow for rapid consumption by those who do not have the time or the expertise to analyse the full set of indicators. In general these methods can be categorized as *numerical aggregation* (e.g. indices), *short selections of indicators* (e.g. core set or headline indicators), *short visual assessments* (e.g. arrows, traffic signals), and *compelling presentations* (e.g. maps or the dashboard of sustainability). Many prominent environmental indicator systems have adjusted their indicator systems to include or report solely on a limited "indicator set" (e.g. the OECD's "Key Environmental Indicators" and the "Canadian Environmental Sustainability Indicators")

Physical Indicators

- Water temperature
- Specific conductance or electrical conductance (EC) or conductivity
- Total suspended solids (TSS)
- Transparency or turbidity
- Total dissolved solids (TDS)
- Odour of water

- Color of water
- Taste of water

Chemical Indicators

- pH
- Biochemical oxygen demand (BOD)
- Chemical oxygen demand (COD)
- Dissolved oxygen (DO)
- Total hardness (TH)
- Heavy metals
- Nitrate
- Orthophosphates
- Pesticides
- Surfactants

Biological indicators

- Ephemeroptera
- Plecoptera
- Mollusca
- Trichoptera
- Escherichia coli (E. coli)
- Coliform bacteria

Biological monitoring metrics have been developed in many places, and one widely used measure is the presence and abundance of members of the insect orders Ephemeroptera, Plecoptera and Trichoptera (common names are, respectively, mayfly, stonefly and caddisfly). EPT indexes will naturally vary from region to region, but generally, within a region, the greater the number of taxa from these orders, the better the water quality. Organisations in the United States, such as EPA offer guidance on developing a monitoring program and identifying members of these and other aquatic insect orders.

Individuals interested in monitoring water quality who cannot afford or manage lab scale analysis can also use biological indicators to get a general reading of water quality. One example is the IOWATER volunteer water monitoring program, which includes a benthic macroinvertebrate indicator key.

Bivalve molluscs are largely used as bioindicators to monitor the health of aquatic environments in both fresh water and the marine environments. Their population status or structure, physiology, behaviour or the level of contamination with elements or compounds can indicate the state of contamination status of the ecosystem. They are particularly useful since they are sessile so that they are representative of the environment where they are sampled or placed. A typical project is the Mussel Watch Programme, but today they are used worldwide.

The Southern African Scoring System (SASS) method is a biological water quality monitoring system based on the presence of benthic macroinvertebrates. The SASS aquatic biomonitoring tool has been refined over the past 30 years and is now on the fifth version (SASS5) which has been specifically modified in accordance with international standards, namely the ISO/IEC 17025 protocol. The SASS5 method is used by the South African Department of Water Affairs as a standard method for River Health Assessment, which feeds the national River Health Programme and the national Rivers Database.

Standards and Reports

International

- The World Health Organisation (WHO) has published guidelines for drinking-water quality (GDWQ) in 2011.
- The International Organization for Standardization (ISO) published regulation of water quality in the section of ICS 13.060, ranging from water sampling, drinking water, industrial class water, sewage, and examination of water for chemical, physical or biological properties. ICS 91.140.60 covers the standards of water supply systems.

National specifications for ambient water and drinking water

European Union

The water policy of the European Union is primarily codified in three directives:

- Directive on Urban Waste Water Treatment (91/271/EEC) of 21 May 1991 concerning discharges of municipal and some industrial wastewaters;
- The Drinking Water Directive (98/83/EC) of 3 November 1998 concerning potable water quality;
- Water Framework Directive (2000/60/EC) of 23 October 2000 concerning water resources management.

India

- Indian Council of Medical Research (ICMR) Standards for Drinking Water.

South Africa

Water quality guidelines for South Africa are grouped according to potential user types (e.g. domestic, industrial) in the 1996 Water Quality Guidelines. Drinking water quality is subject to the South African National Standard (SANS) 241 Drinking Water Specification.

United Kingdom

In England and Wales acceptable levels for drinking water supply are listed in the "Water Supply (Water Quality) Regulations 2000."

United States

In the United States, Water Quality Standards are defined by state agencies for various water bodies, guided by the desired uses for the water body (e.g., fish habitat, drinking water supply, recreational use). The Clean Water Act (CWA) requires each governing jurisdiction (states, territories, and covered tribal entities) to submit a set of biennial reports on the quality of water in their area. These reports are known as the 303(d) and 305(b) reports, named for their respective CWA provisions, and are submitted to, and approved by, EPA. These reports are completed by the governing jurisdiction, typically a state environmental agency. EPA recommends that each state submit a single "Integrated Report" comprising its list of impaired waters and the status of all water bodies in the state. The *National Water Quality Inventory Report to Congress* is a general report on water quality, providing overall information about the number of miles of streams and rivers and their aggregate condition. The CWA requires states to adopt standards for each of the possible designated uses that they assign to their waters. Should evidence suggest or document that a stream, river or lake has failed to meet the water quality criteria for one or more of its designated uses, it is placed on a list of impaired waters. Once a state has placed a water body on this list, it must develop a management plan establishing Total Maximum Daily Loads (TMDLs) for the pollutant(s) impairing the use of the water. These TMDLs establish the reductions needed to fully support the designated uses.

Drinking water standards, which are applicable to public water systems, are issued by EPA under the Safe Drinking Water Act.

WASTEWATER QUALITY INDICATORS

Wastewater quality indicators are laboratory test methodologies to assess suitability of wastewater for disposal or re-use. Tests selected and desired test results vary with the intended use or discharge location. Tests measure physical, chemical, and biological characteristics of the waste water.

Physical Characteristics

Temperature

Aquatic organisms cannot survive outside of specific temperature ranges. Irrigation runoff and water cooling of power stations may elevate temperatures above the acceptable range for some species. **Temperature** may be measured with a calibrated thermometer.

Solids

Solid material in wastewater may be dissolved, suspended, or settled. **Total dissolved solids** or TDS (sometimes called filterable residue) is measured as the mass of residue remaining when a measured volume of filtered water is evaporated. The mass of dried solids remaining on the filter is called **total suspended solids** (TSS) or nonfiltrable residue. Settleable solids are measured as the visible volume accumulated at the bottom of an Imhoff cone after water has settled for one hour. Turbidity is a measure of the light scattering ability of suspended matter in the water. Salinity measures water density or conductivity changes caused by dissolved materials.

Chemical Characteristics

Virtually any chemical may be found in water, but routine testing is commonly limited to a few chemical elements of unique significance.

Hydrogen

Water ionizes into hydronium (H_3O) cations and hydroxyl (OH) anions. The concentration of ionized hydrogen (as protonated water) is expressed as **pH**.

Oxygen

Most aquatic habitats are occupied by fish or other animals requiring certain minimum dissolved oxygen concentrations to survive. Dissolved oxygen concentrations may be measured directly in wastewater, but the amount of oxygen potentially required by other chemicals in the wastewater is termed an oxygen demand. Dissolved or suspended oxidizable organic material in wastewater will be used as a food source. Finely divided material is readily available to microorganisms whose populations will increase to digest the amount of food available. Digestion of this food requires oxygen, so the oxygen content of the water will ultimately be decreased by the amount required to digest the dissolved or suspended food. Oxygen concentrations may fall below the minimum required by aquatic animals if the rate of oxygen utilization exceeds replacement by atmospheric oxygen.

The reaction for biochemical oxidation may be written as:

Oxidizable material + bacteria + nutrient + $O_2 \rightarrow CO_2 + H_2O$ + oxidized inorganics such as NO_3 or SO_4

Oxygen consumption by reducing chemicals such as sulfides and nitrites is typified as follows:

$$S^{2-} + 2\,O_2 \rightarrow SO_4^{2-}$$

$$NO_2^- + \tfrac{1}{2}\,O_2 \rightarrow NO_3^-$$

Since all natural waterways contain bacteria and nutrient, almost any waste compounds introduced into such waterways will initiate biochemical reactions (such as shown above).

Those biochemical reactions create what is measured in the laboratory as the biochemical oxygen demand (BOD).

Oxidizable chemicals (such as reducing chemicals) introduced into a natural water will similarly initiate chemical reactions (such as shown above). Those chemical reactions create what is measured in the laboratory as the chemical oxygen demand (COD).

Both the BOD and COD tests are a measure of the relative oxygen-depletion effect of a waste contaminant. Both have been widely adopted as a measure of pollution effect. The BOD test measures the oxygen demand of biodegradable pollutants whereas the COD test measures the oxygen demand of biodegradable pollutants plus the oxygen demand of non-biodegradable oxidizable pollutants.

The so-called 5-day BOD measures the amount of oxygen consumed by biochemical oxidation of waste contaminants in a 5-day period. The total amount of oxygen consumed when the biochemical reaction is allowed to proceed to completion is called the Ultimate BOD. The Ultimate BOD is too time consuming, so the 5-day BOD has almost universally been adopted as a measure of relative pollution effect.

There are also many different COD tests. Perhaps, the most common is the 4-hour COD.

There is no generalized correlation between the 5-day BOD and the Ultimate BOD. Likewise, there is no generalized correlation between BOD and COD. It is possible to develop such correlations for a specific waste contaminant in a specific wastewater stream, but such correlations cannot be generalized for use with any other waste contaminants or wastewater streams.

The laboratory test procedures for determining the above oxygen demands are detailed in the following sections of the "Standard Methods For the Examination Of Water and Wastewater" available at www.standardmethods.org:

- 5-day BOD and Ultimate BOD: Sections 5210B and 5210C
- COD: Section 5220

Nitrogen

Nitrogen is an important nutrient for plant and animal growth. Atmospheric nitrogen is less biologically available than dissolved nitrogen in the form of ammonia and nitrates. Availability of dissolved nitrogen may contribute to algal blooms. Ammonia and organic forms of nitrogen are often measured as **Total Kjeldahl Nitrogen**, and analysis for inorganic forms of nitrogen may be performed for more accurate estimates of total nitrogen content.

Phosphates

Total Phosphorus and Phosphate, PO^{-3}_{4}

Phosphates enter the water ways through both non-point sources and point sources. Non-point source (NPS) pollution refers to water pollution from diffuse sources. Nonpoint source pollution can be contrasted with point source pollution, where discharges occur to a body of water at a single location. The non-point sources of phosphates include: natural decomposition of rocks and minerals, storm water runoff, agricultural runoff, erosion and sedimentation, atmospheric deposition, and direct input by animals/wildlife; whereas: point sources may include: waste water treatment plants and permitted industrial discharges. In general, the non-point source pollution typically is significantly higher than the point sources of pollution. Therefore, the key to sound management is to limit the input from both point and non-point sources of phosphate. High concentration of phosphate in water bodies is an indication of pollution and largely responsible for eutrophication.

Phosphates are not toxic to people or animals unless they are present in very high levels. Digestive problems could occur from extremely high levels of phosphate.

The following criteria for total phosphorus were recommended by the U.S. Environmental Protection Agency.

- No more than 0.1 mg/L for streams which do not empty into reservoirs,
- No more than 0.05 mg/L for streams discharging into reservoirs, and
- No more than 0.025 mg/L for reservoirs.

Phosphorus is normally low (< 1 mg/l) in clean potable water sources and usually not regulated;

Chlorine

Chlorine has been widely used for bleaching, as a disinfectant, and for biofouling prevention in water cooling systems. Remaining concentrations of oxidizing hypochlorous acid and hypochlorite ions may be measured as chlorine residual to estimate effectiveness of disinfection or to demonstrate safety for discharge to aquatic ecosystems.

Biological Characteristics

Water may be tested by a bioassay comparing survival of an aquatic test species in the wastewater in comparison to water from some other source. Water may also be evaluated to determine the approximate biological population of the wastewater. Pathogenic micro-organisms using water as a means of moving from one host to another may be present in sewage. Coliform index measures the population of an organism commonly found in the intestines of warm-blooded animals as an indicator of the possible presence of other intestinal pathogens.

WATER TREATMENT

Water treatment is any process that makes water more acceptable for a specific end-use. The end use may be drinking, industrial water supply, irrigation, river flow maintenance, water recreation or many other uses, including being safely returned to the environment. Water treatment removes contaminants and undesirable components, or reduces their concentration so that the water becomes fit for its desired end-use.

Abandoned Water Purification Plant Springfield, Tennessee, United States

Treatment for Drinking Water Production

Treatment for drinking water production involves the removal of contaminants from raw water to produce water that is pure enough for human consumption without any short term or long term risk of any adverse health effect. Substances that are removed during the process of drinking water treatment include suspended solids, bacteria, algae, viruses, fungi, and minerals such as iron and manganese.

The processes involved in removing the contaminants include physical processes such as settling and filtration, chemical processes such as disinfection and coagulation and biological processes such as slow sand filtration.

Measures taken to ensure water quality not only relate to the treatment of the water, but to its conveyance and distribution after treatment. It is therefore common practice to keep residual disinfectants in the treated water to kill bacteriological contamination during distribution.

World Health Organization (WHO) guidelines are a general set of standards intended to apply where better local standards are not implemeted. More rigorous standards apply across Europe, the USA and in most other developed countries. followed throughout the world for drinking water quality requirements.

Processes

A combination selected from the following processes is used for municipal drinking water treatment worldwide:

- Pre-chlorination for algae control and arresting biological growth
- Aeration along with pre-chlorination for removal of dissolved iron when present with small amouts relatively of manganese
- Coagulation for flocculation or slow-sand filtration

- Coagulant aids, also known as polyelectrolytes – to improve coagulation and for more robust floc formation
- Sedimentation for solids separation that is removal of suspended solids trapped in the floc
- Filtration to remove particles from water either either by passage through a sand bed that can be washed and reused or by passage through a purpose designed filter that may be washable.
- Disinfection for killing bacteria viruses and other pathogens.

Empty aeration tank for iron precipitation

Tanks with sand filters to remove precipitated iron (not working at the time)

Technologies for potable water and other uses are well developed, and generalized designs are available from which treatment processes can be selected for pilot testing on the specific source water. In addition, a number of private companies provide patented technological solutions for treatment of specific contaminants. Automation of water and waste-water treatment is common in the developed world. Source water quality through the seasons, scale and environmental impact can dictate capital costs and operating costs. End use of the treated water dictates the necessary quality monitoring technologies, and locally available skills typically dictate the level of automation adopted.

Polluted Water Treatment

Wastewater treatment is the process that removes the majority of the contaminants from wastewater or sewage and produces both a liquid effluent suitable for disposal to the natural environment and a sludge. Biological processes can be employed in the treatment of wastewater and these processes may include, for example, aerated lagoons, activated sludge or slow sand filters.

A sewage treatment plant in northern Portugal.

To be effective, sewage must be conveyed to a treatment plant by appropriate pipes and infrastructure and the process itself must be subject to regulation and controls. Some wastewaters require different and sometimes specialized treatment methods. At the simplest level, treatment of sewage and most wastewaters is carried out through separation of solids from liquids, usually by sedimentation. By progressively converting dissolved material into solids, usually a biological floc, which is then settled out, an effluent stream of increasing purity is produced.

Industrial Water and Wastewater Treatment

Two of the main processes of industrial water treatment are *boiler water treatment* and *cooling water treatment*. A lack of proper water treatment can lead to the reaction of solids and bacteria within pipe work and boiler housing. Steam boilers can suffer from scale or corrosion when left untreated. Scale deposits can lead to weak and dangerous machinery, while additional fuel is required to heat the same level of water because of the rise in thermal resistance. Poor quality dirty water can become a breeding ground for bacteria such as *Legionella* causing a risk to public health.

With the proper treatment, a significant proportion of industrial on-site wastewater might be reusable. This can save money in three ways: lower charges for lower water consumption, lower charges for the smaller volume of effluent water discharged and lower energy costs due to the recovery of heat in recycled wastewater.

Corrosion in low pressure boilers can be caused by dissolved oxygen, acidity and excessive alkalinity. Water treatment therefore should remove the dissolved oxygen and maintain the boiler water with the appropriate pH and alkalinity levels. Without effective water treatment, a cooling water system can suffer from scale formation, corrosion and fouling and may become a breeding ground for harmful bacteria. This reduces efficiency, shortens plant life and makes operations unreliable and unsafe.

Domestic Water Treatment

Water supplied to domestic properties may be further treated before use, often using an in-line treatment process. Such treatments can include water softening or ion exchange. Many propriety systems also claim to remove residual disinfectants and heavy metal ions.

Desalination

Saline water can be treated to yield fresh water. Two main processes are used, reverse osmosis or distillation. Both methods require high energy inputs and are usually only used where fresh water is difficult to source.

Field Processes

Living away from drinking water supplies often requires some form of portable water treatment process. These can vary in complexity from the simple addition of a disinfectant tablet in a hiker's water bottle through to complex multi-stage processes carried by boat or plane to disaster areas.

Ultra Pure Water Production

Some industries such as the production of silicon wafers, space technology and many high quality metallurgical process require ultrapure water. The production of such water typically involves many stages, and can include reverse osmosis, ion exchange and several distillation stages using solid tin apparatus.

History

Early water treatment methods still used included sand filtration and chlorination. The first documented use of sand filters to purify the water supply dates to 1804, when the owner of a bleachery in Paisley, Scotland, John Gibb, installed an experimental filter, selling his unwanted surplus to the public. This method was refined in the following two decades, and it culminated in the first treated public water supply in the world, installed by the Chelsea Waterworks Company in London in 1829.

Society and Culture

Developing Countries

As of 2006, waterborne diseases are estimated to have caused 1.8 million deaths each year. These deaths are attributable to inadequate public sanitation systems and in these cases, proper sewerage (or other options such as small-scale wastewater treatment) that must be installed.

Appropriate technology options in water treatment include both community-scale and household-scale point-of-use (POU) designs. Such designs may employ solar water disinfection methods, using solar irradiation to inactivate harmful waterborne microorganisms directly, mainly by the UV-A component of the solar spectrum, or indirectly through the presence of an oxide photocatalyst, typically supported TiO_2 in its anatase or rutile phases. Despite progress in SODIS technology, military surplus water treatment units like the ERDLator are still frequently used in developing countries. Newer military style Reverse Osmosis Water Purification Units (ROWPU) are portable, self-contained water treatment plants are becoming more available for public use.

For waterborne disease reduction to last, water treatment programs that research and development groups start in developing countries must be sustainable by the citizens of those countries. This can ensure the efficiency of such programs after the departure of the research team, as monitoring is difficult because of the remoteness of many locations.

Industrial Water Treatment

There are many uses of water in industry and,in most cases, the used water also needs treatment to render it fit for re-use or disposal. Raw water entering an industrial plant often needs treatment to meet tight quality specifications to be of use in specific industrial processes. **Industrial water treatment** encompasses all these aspects which include industrial wastewater treatment, boiler water treatment and cooling water treatment.

Overview

Water treatment is used to optimize most water-based industrial processes, such as heating, cooling, processing, cleaning, and rinsing so that operating costs and risks are reduced. Poor water treatment lets water interact with the surfaces of pipes and vessels which contain it. Steam boilers can scale up or corrode, and these deposits will mean more fuel is needed to heat the same amount of water. Cooling towers can also scale up and corrode, but left untreated, the warm, dirty water they can contain will encourage bacteria to grow, and Legionnaires' disease can be the fatal consequence. Water treatment is also used to improve the quality of water contacting the manufactured product e.g. semiconductors, and/or can be part of the product e.g. beverages, pharmaceuticals, etc. In these instances, poor water treatment can cause defective products.

In many cases, effluent water from one process can be suitable for reuse in another process if given suitable treatment. This can reduce costs by lowering charges for water consumption, reduce the costs of effluent disposal because of reduced volume and lower energy costs due to the recovery of heat in recycled wastewater.

Objectives

Industrial water treatment seeks to manage four main problem areas: scaling, corrosion, microbiological activity and disposal of residual wastewater. Boilers do not have many problems with microbes as the high temperatures prevent their growth.

Scaling occurs when the chemistry and temperature conditions are such that the dissolved mineral salts in the water are caused to precipitate and form solid deposits. These can be mobile, like a fine silt, or can build up in layers on the metal surfaces of the systems. Scale is a problem because it insulates and heat exchange becomes less efficient as the scale thickens, which wastes energy. Scale also narrows pipe widths and therefore increases the energy used in pumping the water through the pipes.

Corrosion occurs when the parent metal oxidises (as iron rusts, for example) and gradually the integrity of the plant equipment is compromised. The corrosion products can cause similar problems to scale, but corrosion can also lead to leaks, which in a pressurised system can lead to catastrophic failures.

Microbes can thrive in untreated cooling water, which is warm and sometimes full of organic nutrients as wet cooling towers are very efficient air scrubbers. Dust, flies, grass, fungal spores, and others collect in the water and create a sort of "microbial soup" if not treated with biocides. Most outbreaks of the deadly Legionnaires' Disease have been traced to unmanaged cooling towers, and the UK has had stringent Health & Safety guidelines concerning cooling tower operations for many years as have had governmental agencies in other countries.

Disposal of residual industrial wastewaters

Disposal of residual wastewaters from an industrial plant is a difficult and costly problem. Most petroleum refineries, chemical and petrochemical plants have onsite facilities to treat their wastewaters so that the pollutant concentrations in the treated wastewater comply with the local and/or national regulations regarding disposal of wastewaters into community treatment plants or into rivers, lakes or oceans.

Technologies

Advancements in water treatment technology have affected all areas of industrial water treatment. Although mechanical filtration, such as reverse osmosis, is widely employed to filter contaminants, other technologies including the use of ozone generators, wastewater evaporation, electrodeionization and bioremediation are also able to address the challenges of industrial water treatment.

Ozone treatment is a process in which ozone gas is injected into waste streams as a means to reduce or eliminate the need for water treatment chemicals or sanitizers that may be hazardous, including chlorine.

Ultraviolet Irradiation

Ultraviolet (UV) disinfection technology has been a common water treatment technology in the past two decades due to its ability to provide disinfected water without the use of harmful chemicals. The UV-C portion represents wavelengths from 200 nm - 280 nm which is used for disinfection. UV-C photons penetrate cells and damage the nucleic acid, rendering them incapable of reproduction, or microbiologically inactive.

Process Water Treatment Technology

Process water is water that is used in a variety of manufacturing operations, such as: coating and plating; rinsing and spraying; washing and etc. Municipal and ground water often contain dissolved minerals which make it unsuitable for these processes because it would affect product quality and/or increase manufacturing costs. A proper incoming water treatment system can remedy these issues and create the right water conditions for specific industrial processes.

Industrial Wastewater Treatment

Industrial wastewater treatment covers the mechanisms and processes used to treat wastewater that is produced as a by-product of industrial or commercial activities. After treatment, the treated industrial wastewater (or effluent) may be reused or released to a sanitary sewer or to a surface water in the environment. Most industries produce some wastewater although recent trends in the developed world have been to minimise such production or recycle such wastewater within the production process. However, many industries remain dependent on processes that produce wastewaters.

Sources of Industrial Wastewater

Complex Organic Chemicals Industry

A range of industries manufacture or use complex organic chemicals. These include pesticides, pharmaceuticals, paints and dyes, petrochemicals, detergents, plastics, paper pollution, etc. Waste waters can be contaminated by feedstock materials, by-products, product material in soluble or particulate form, washing and cleaning agents, solvents and added value products such as plasticisers. Treatment facilities that do not need control of their effluent typically opt for a type of aerobic treatment, i.e. aerated lagoons.

Electric Power Plants

Fossil-fuel power stations, particularly coal-fired plants, are a major source of industrial wastewater. Many of these plants discharge wastewater with significant levels of metals such as lead, mercury, cadmium and chromium, as well as arsenic, selenium and nitrogen compounds (nitrates and nitrites). Wastewater streams include flue-gas desulfurization, fly ash, bottom ash and flue gas mercury control. Plants with air pollution controls such as wet scrubbers typically transfer the captured pollutants to the wastewater stream.

Ash ponds, a type of surface impoundment, are a widely used treatment technology at coal-fired plants. These ponds use gravity to settle out large particulates (measured as total suspended solids) from power plant wastewater. This technology does not treat dissolved pollutants. Power stations use additional technologies to control pollutants, depending on the particular wastestream in the plant. These include dry ash handling, closed-loop ash recycling, chemical precipitation, biological treatment (such as an activated sludge process), and evaporation.

Food Industry

Wastewater generated from agricultural and food operations has distinctive characteristics that set it apart from common municipal wastewater managed by public or private sewage treatment plants throughout the world: it is biodegradable and non-toxic, but has high concentrations of biochemical oxygen demand (BOD) and suspended solids (SS). The constituents of food and agriculture wastewater are often complex to predict, due to the differences in BOD and pH in effluents from vegetable, fruit, and meat products and due to the seasonal nature of food processing and post-harvesting.

Processing of food from raw materials requires large volumes of high grade water. Vegetable washing generates waters with high loads of particulate matter and some dissolved organic matter. It may also contain surfactants.

Animal slaughter and processing produces very strong organic waste from body fluids, such as blood, and gut contents. This wastewater is frequently contaminated by significant levels of antibiotics and growth hormones from the animals and by a variety of pesticides used to control external parasites.

Processing food for sale produces wastes generated from cooking which are often rich in plant organic material and may also contain salt, flavourings, colouring material and acids or alkali. Very significant quantities of oil or fats may also be present.

Iron and Steel Industry

The production of iron from its ores involves powerful reduction reactions in blast furnaces. Cooling waters are inevitably contaminated with products especially ammonia and cyanide. Production of coke from coal in coking plants also requires water cooling and the use of water in by-products separation. Contamination of waste streams includes gasification products such as benzene, naphthalene, anthracene, cyanide, ammonia, phenols, cresols together with a range of more complex organic compounds known collectively as polycyclic aromatic hydrocarbons (PAH).

The conversion of iron or steel into sheet, wire or rods requires hot and cold mechanical transformation stages frequently employing water as a lubricant and coolant. Contaminants include hydraulic oils, tallow and particulate solids. Final treatment of iron and steel products before onward sale into manufacturing includes *pickling* in strong mineral acid to remove rust and prepare the surface for tin or chromium plating or for other surface treatments such as galvanisation or painting. The two acids commonly used are hydrochloric acid and sulfuric acid. Wastewaters include acidic rinse waters together with waste acid. Although many plants operate acid recovery plants (particularly those using hydrochloric acid), where the mineral acid is boiled away from the iron salts, there remains a large volume of highly acid ferrous sulfate or ferrous chloride to be disposed of. Many steel industry wastewaters are contaminated by hydraulic oil, also known as *soluble oil.*

Mines and Quarries

The principal waste-waters associated with mines and quarries are slurries of rock particles in water. These arise from rainfall washing exposed surfaces and haul roads and also from rock washing and grading processes. Volumes of water can be very high, especially rainfall related arisings on large sites. Some specialized separation operations, such as coal washing to separate coal from native rock using density gradients, can produce wastewater contaminated by fine particulate haematite and surfactants. Oils and hydraulic oils are also common contaminants.

Mine wastewater effluent in Peru, with neutralized pH from tailing runoff.

Wastewater from metal mines and ore recovery plants are inevitably contaminated by the minerals present in the native rock formations. Following crushing and extraction of the desirable materials, undesirable materials may enter the wastewater stream. For metal mines, this can include unwanted metals such as zinc and other materials such as arsenic. Extraction of high value metals such as gold and silver may generate slimes containing very fine particles in where physical removal of contaminants becomes particularly difficult.

Additionally, the geologic formations that harbour economically valuable metals such as copper and gold very often consist of sulphide-type ores. The processing entails grinding the rock into fine particles and then extracting the desired metal(s), with the leftover rock being known as tailings. These tailings contain a combination of not only undesirable leftover metals, but also sulphide components which eventually form sulphuric acid upon the exposure to air and water that inevitably occurs when the tailings are disposed of in large impoundments. The resulting acid mine drainage, which is often rich in heavy metals (because acids dissolve metals), is one of the many environmental impacts of mining.

Nuclear Industry

The waste production from the nuclear and radio-chemicals industry is dealt with as *Radioactive waste.*

Pulp and Paper Industry

Effluent from the pulp and paper industry is generally high in suspended solids and BOD. Plants that bleach wood pulp for paper making may generate chloroform, dioxins (including 2,3,7,8-TCDD), furans, phenols and chemical oxygen demand (COD). Stand-alone paper mills using imported pulp may only require simple primary treatment, such as sedimentation or dissolved air flotation. Increased BOD or COD loadings, as well as organic pollutants, may require biological treatment such as activated sludge or upflow anaerobic sludge blanket reactors. For mills with high inorganic loadings like salt, tertiary treatments may be required, either general membrane treatments like ultrafiltration or reverse osmosis or treatments to remove specific contaminants, such as nutrients.

Industrial Oil Contamination

Industrial applications where oil enters the wastewater stream may include vehicle wash bays, workshops, fuel storage depots, transport hubs and power generation. Often the wastewater is discharged into local sewer or trade waste systems and must meet local environmental specifications. Typical contaminants can include solvents, detergents, grit. lubricants and hydrocarbons.

Water Treatment

Many industries have a need to treat water to obtain very high quality water for demanding purposes such as environmental discharge compliance. Water treatment produces organic and mineral sludges from filtration and sedimentation. Ion exchange using natural or synthetic resins removes calcium, magnesium and carbonate ions from water, typically replacing them with sodium, chloride, hydroxyl and/or other ions. Regeneration of ion exchange columns with strong acids and alkalis produces a wastewater rich in hardness ions which are readily precipitated out, especially when in admixture with other wastewater constituents.

Wool processing

Insecticide residues in fleeces are a particular problem in treating waters generated in wool processing. Animal fats may be present in the wastewater, which if not contaminated, can be recovered for the production of tallow or further rendering.

Treatment of Industrial Wastewater

The various types of contamination of wastewater require a variety of strategies to remove the contamination.

Brine Treatment

Brine treatment involves removing dissolved salt ions from the waste stream. Although similarities to seawater or brackish water desalination exist, industrial brine treatment may contain unique combinations of dissolved ions, such as hardness ions or other metals, necessitating specific processes and equipment.

Brine treatment systems are typically optimized to either reduce the volume of the final discharge for more economic disposal (as disposal costs are often based on volume) or maximize the recovery of fresh water or salts. Brine treatment systems may also be optimized to reduce electricity consumption, chemical usage, or physical footprint.

Brine treatment is commonly encountered when treating cooling tower blowdown, produced water from steam assisted gravity drainage (SAGD), produced water from natural gas extraction such as coal seam gas, frac flowback water, acid mine or acid rock drainage, reverse osmosis reject, chlor-alkali wastewater, pulp and paper mill effluent, and waste streams from food and beverage processing.

Brine treatment technologies may include: membrane filtration processes, such as reverse osmosis; ion exchange processes such as electrodialysis or weak acid cation exchange; or evaporation processes, such as brine concentrators and crystallizers employing mechanical vapour recompression and steam.

Reverse osmosis may not be viable for brine treatment, due to the potential for fouling caused by hardness salts or organic contaminants, or damage to the reverse osmosis membranes from hydrocarbons.

Evaporation processes are the most widespread for brine treatment as they enable the highest degree of concentration, as high as solid salt. They also produce the highest purity effluent, even distillate-quality. Evaporation processes are also more tolerant of organics, hydrocarbons, or hardness salts. However, energy consumption is high and corrosion may be an issue as the prime mover is concentrated salt water. As a result, evaporation systems typically employ titanium or duplex stainless steel materials.

Brine Management

Brine management examines the broader context of brine treatment and may include consideration of government policy and regulations, corporate sustainability, environmental impact, recycling, handling and transport, containment, centralized compared to on-site treatment, avoidance and reduction, technologies, and economics. Brine management shares some issues with leachate management and more general waste management.

Solids Removal

Most solids can be removed using simple sedimentation techniques with the solids recovered as slurry or sludge. Very fine solids and solids with densities close to the density of water pose special problems. In such case filtration or ultrafiltration may be required. Although, flocculation may be used, using alum salts or the addition of polyelectrolytes.

Oils and Grease Removal

The effective removal of oils and grease is dependent on the characteristics of the oil in terms of its suspension state and droplet size, which will in turn affect the choice of separator technology.

Oil pollution in water usually comes in four states, often in combination:

- free oil - large oil droplets sitting on the surface;
- heavy oil, which sits at the bottom, often adhering to solids like dirt;
- emulsified, where the oil droplets are heavily "chopped"; and
- dissolved oil, where the droplets are fully dispersed and not visible. Emulsified oil droplets are the most common in industrial oily wastewater and are extremely difficult to separate.

The methodology for separating the oil is dependent on the oil droplet size. Larger oil droplets such as those in free oil pollution are easily removed, but as the droplets become smaller, some separator technologies perform better than others.

Most separator technologies will have an optimum range of oil droplet sizes that can be effectively treated. This is known as the "micron rating."

Analysing the oily water to determine droplet size can be performed with a video particle analyser. Alternatively, there are commonalities in industries for oil droplet sizes. Larger droplets–greater than 60 microns–are often present in wastewater in workshops, re-fuel areas and depots. Twenty to 50 micron oil droplets often are present in vehicle wash bays, meat processing and dairy manufacturing effluent and aluminium billet cooling towers. Smaller droplets in the range of 10 to 20 microns tend to occur in workshops and condensates.

Each separator technology will have its' own performance curve outlining optimum performance based on oil droplet size. the most common separators are gravity tanks or pits, API oil-water separators or plate packs, chemical treatment via DAFs, centrifuges, media filters and hydrocyclones.

API separators

Many oils can be recovered from open water surfaces by skimming devices. Considered a dependable and cheap way to remove oil, grease and other hydrocarbons from water, oil skimmers can sometimes achieve the desired level of water purity. At other times, skimming is also a cost-efficient method to remove most of the oil before using membrane filters and chemical processes. Skimmers will prevent filters from blinding prematurely and keep chemical costs down because there is less oil to process.

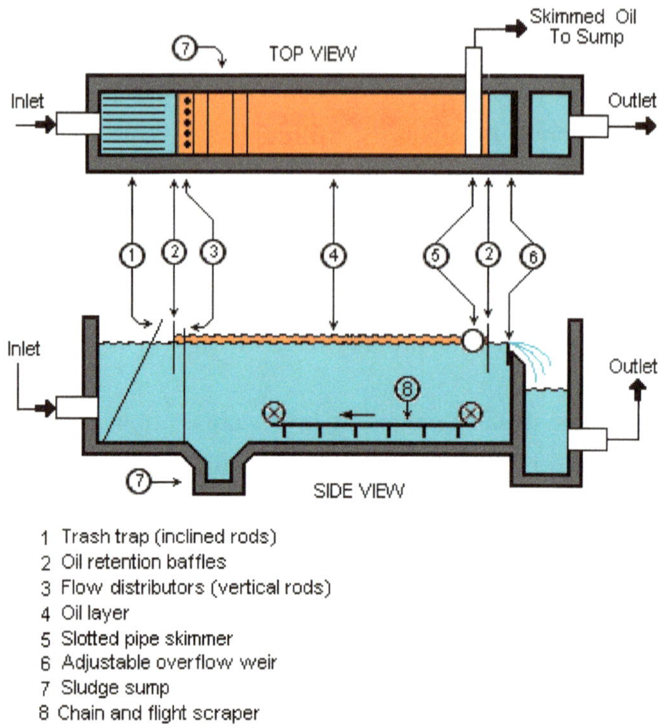

A typical API oil-water separator used in many industries

Because grease skimming involves higher viscosity hydrocarbons, skimmers must be equipped with heaters powerful enough to keep grease fluid for discharge. If floating grease forms into solid clumps or mats, a spray bar, aerator or mechanical apparatus can be used to facilitate removal.

However, hydraulic oils and the majority of oils that have degraded to any extent will also have a soluble or emulsified component that will require further treatment to eliminate. Dissolving or emulsifying oil using surfactants or solvents usually exacerbates the problem rather than solving it, producing wastewater that is more difficult to treat.

The wastewaters from large-scale industries such as oil refineries, petrochemical plants, chemical plants, and natural gas processing plants commonly contain gross amounts of oil and suspended solids. Those industries use a device known as an API oil-water separator which is designed to separate the oil and suspended solids from their wastewater effluents. The name is derived from the fact that such separators are designed according to standards published by the American Petroleum Institute (API).

The API separator is a gravity separation device designed by using Stokes Law to define the rise velocity of oil droplets based on their density and size. The design is based on the specific gravity difference between the oil and the wastewater because that difference is much smaller than the specific gravity difference between the suspended solids and water. The suspended solids settles to the bottom of the separator as a sediment layer, the oil rises to top of the separator and the cleansed wastewater is the middle layer between the oil layer and the solids.

Typically, the oil layer is skimmed off and subsequently re-processed or disposed of, and the bottom sediment layer is removed by a chain and flight scraper (or similar device) and a sludge pump. The water layer is sent to further treatment for additional removal of any residual oil and then to some type of biological treatment unit for removal of undesirable dissolved chemical compounds.

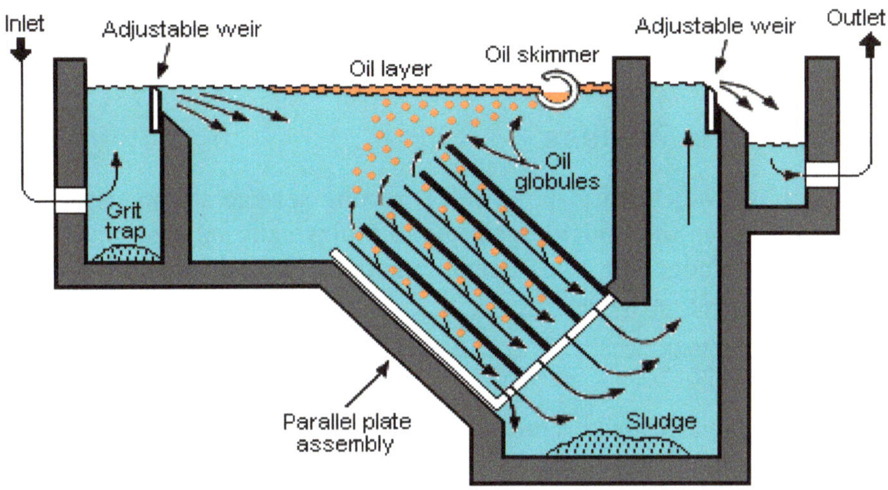

A typical parallel plate separator

Parallel plate separators are similar to API separators but they include tilted parallel plate assemblies (also known as parallel packs). The parallel plates provide more surface for suspended oil droplets to coalesce into larger globules. Such separators still depend upon the specific gravity between the suspended oil and the water. However, the parallel plates enhance the degree of oil-water separation. The result is that a parallel plate separator requires significantly less space than a conventional API separator to achieve the same degree of separation.

Hydrocyclone Oil Separators

Hydrocyclone oil separators operate on the process where wastewater enters the cyclone chamber and is spun under extreme centrifugal forces more than 1000 times the force of gravity. This force causes the water and oil droplets to separate. The separated oil is discharged from one end of the cyclone where treated water is discharged through the opposite end for further treatment, filtration or discharge.

Hydrocyclones are useful for the greatest range of oil droplet sizes operating from less than 10 microns and up and can operate continuously without water pre-treatment and at any temperature and pH. Applications where hydrocyclones are found are in industry where oily water sources arise in workshops, vehicle wash bays, transport hubs, fuel depots and aluminium billet processing. Animal fats from meat processing and dairy manufacturing can also be removed without the need of chemical treatment that often is required for dissolved air flotation (DAF) systems.

Removal of Biodegradable Organics

Biodegradable organic material of plant or animal origin is usually possible to treat using extended conventional sewage treatment processes such as activated sludge or trickling filter. Problems can arise if the wastewater is excessively diluted with washing water or is highly concentrated such as undiluted blood or milk. The presence of cleaning agents, disinfectants, pesticides, or antibiotics can have detrimental impacts on treatment processes.

Activated Sludge Process

Activated sludge is a biochemical process for treating sewage and industrial wastewater that uses air (or oxygen) and microorganisms to biologically oxidize organic pollutants, producing a waste sludge (or floc) containing the oxidized material. In general, an activated sludge process includes:

- An aeration tank where air (or oxygen) is injected and thoroughly mixed into the wastewater.
- A settling tank (usually referred to as a clarifier or "settler") to allow the waste sludge to settle. Part of the waste sludge is recycled to the aeration tank and the remaining waste sludge is removed for further treatment and ultimate disposal.

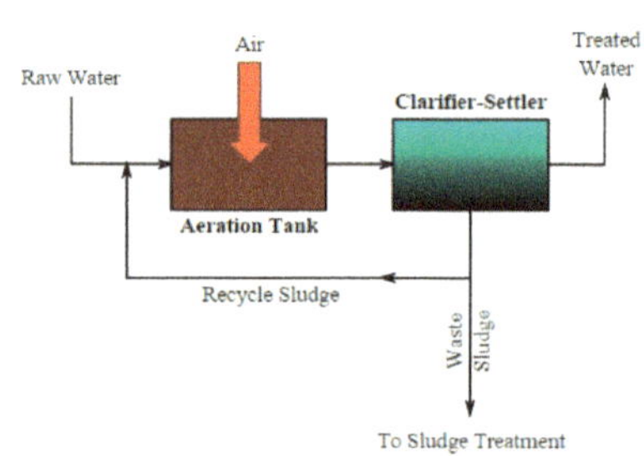

A generalized diagram of an activated sludge process.

Trickling Filter Process

A trickling filter consists of a bed of rocks, gravel, slag, peat moss, or plastic media over which wastewater flows downward and contacts a layer (or film) of microbial slime covering the bed media. Aerobic conditions are maintained by forced air flowing through the bed or by natural convection of air. The process involves adsorption of organic compounds in the wastewater by the microbial slime layer, diffusion of air into the slime layer to provide the oxygen required for the biochemical oxidation of the organic compounds.

The end products include carbon dioxide gas, water and other products of the oxidation. As the slime layer thickens, it becomes difficult for the air to penetrate the layer and an inner anaerobic layer is formed.

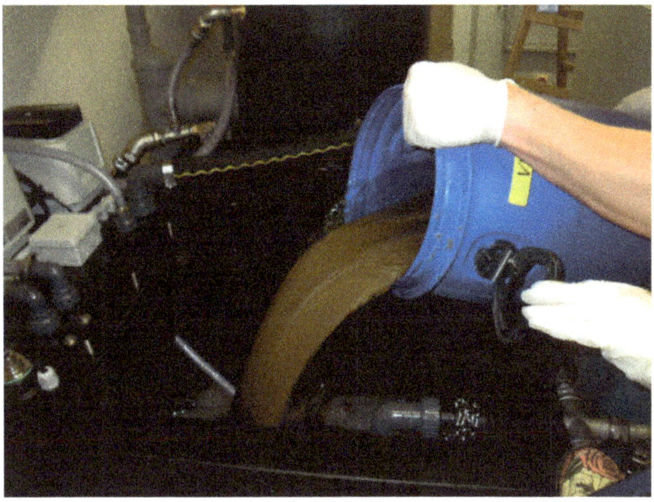

A schematic cross-section of the contact face of the bed media in a trickling filter

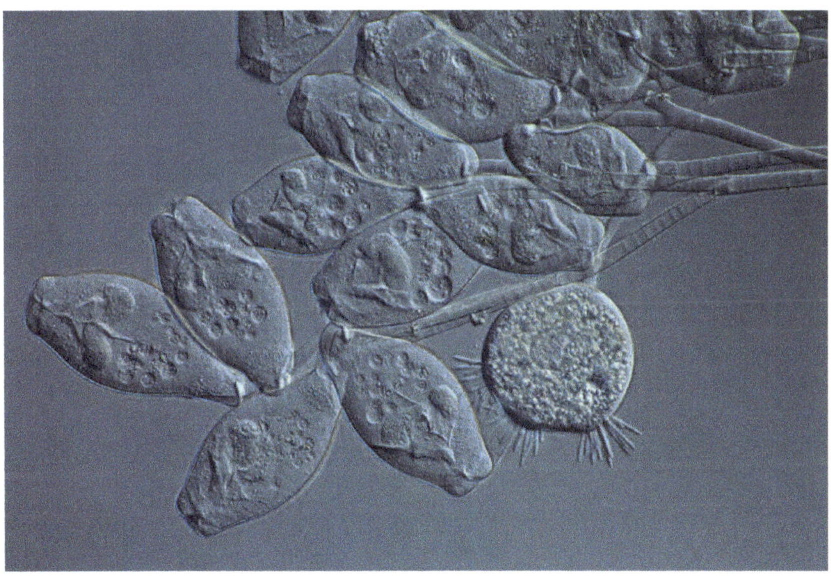

A typical complete trickling filter system

The fundamental components of a complete trickling filter system are:

- A bed of filter medium upon which a layer of microbial slime is promoted and developed.

- An enclosure or a container which houses the bed of filter medium.
- A system for distributing the flow of wastewater over the filter medium.
- A system for removing and disposing of any sludge from the treated effluent.

The treatment of sewage or other wastewater with trickling filters is among the oldest and most well characterized treatment technologies.

A trickling filter is also often called a *trickle filter*, *trickling biofilter*, *biofilter*, *biological filter* or *biological trickling filter*.

Treatment of Other Organics

Synthetic organic materials including solvents, paints, pharmaceuticals, pesticides, products from coke production and so forth can be very difficult to treat. Treatment methods are often specific to the material being treated. Methods include advanced oxidation processing, distillation, adsorption, vitrification, incineration, chemical immobilisation or landfill disposal. Some materials such as some detergents may be capable of biological degradation and in such cases, a modified form of wastewater treatment can be used.

Treatment of Acids and Alkalis

Acids and alkalis can usually be neutralised under controlled conditions. Neutralisation frequently produces a precipitate that will require treatment as a solid residue that may also be toxic. In some cases, gases may be evolved requiring treatment for the gas stream. Some other forms of treatment are usually required following neutralisation.

Waste streams rich in hardness ions as from de-ionisation processes can readily lose the hardness ions in a buildup of precipitated calcium and magnesium salts. This precipitation process can cause severe *furring* of pipes and can, in extreme cases, cause the blockage of disposal pipes. A 1 metre diameter industrial marine discharge pipe serving a major chemicals complex was blocked by such salts in the 1970s. Treatment is by concentration of de-ionisation waste waters and disposal to landfill or by careful pH management of the released wastewater.

Treatment of Toxic Materials

Toxic materials including many organic materials, metals (such as zinc, silver, cadmium, thallium, etc.) acids, alkalis, non-metallic elements (such as arsenic or selenium) are generally resistant to biological processes unless very dilute. Metals can often be precipitated out by changing the pH or by treatment with other chemicals. Many, however, are resistant to treatment or mitigation and may require concentration followed by landfilling or recycling. Dissolved organics can be *incinerated* within the wastewater by the advanced oxidation process.

REFERENCES

1. Diersing, Nancy (2009). "Water Quality: Frequently Asked Questions." Florida Brooks National Marine Sanctuary, Key West, FL.

2. Johnson, D.L., S.H. Ambrose, T.J. Bassett, M.L. Bowen, D.E. Crummey, J.S. Isaacson, D.N. Johnson, P. Lamb, M. Saul, and A.E. Winter-Nelson (1997). "Meanings of environmental terms." Journal of Environmental Quality. 26: 581-589. doi:10.2134/jeq1997.00472425002600030002x
3. "What are Water Quality Standards?". Washington, D.C.: U.S. Environmental Protection Agency (EPA). 2016-03-17.
4. Babbitt, Harold E. & Doland, James J. Water Supply Engineering (1949) McGraw-Hill p.388
5. Linsley, Ray K. & Franzini, Joseph B. Water-Resources Engineering (1972) McGraw-Hill ISBN 0-07-037959-9 pp.454-456
6. World Health Organization (2004). "Consensus of the Meeting: Nutrient minerals in drinking-water and the potential health consequences of long-term consumption of demineralized and remineralized and altered mineral content drinking-waters." Rolling Revision of the WHO Guidelines for Drinking-Water Quality (draft). From November 11–13, 2003 meeting in Rome, Italy at the WHO European Centre for Environment and Health.
7. Canencia, Oliva P; Dalugdug, Marlou D; Emano, Athena Marie; Mendoza, Richard; Walag, Angelo Mark P. (2016-08-31). "Slaughter waste effluents and river catchment watershed contamination in Cagayan de Oro City, Philippines". ResearchGate. 9 (2). ISSN 2220-6663.
8. Goldman, Charles R. & Horne, Alexander J. Limnology (1983) McGraw-Hill ISBN 0-07-023651-8 chapter 6
9. Franson, Mary Ann (1975). Standard Methods for the Examination of Water and Wastewater 14th ed. Washington, DC: American Public Health Association, American Water Works Association & Water Pollution Control Federation. ISBN 0-87553-078-8
10. "Chapter 8. Data Analysis". Handbook for Monitoring Industrial Wastewater (Report). EPA. August 1973. EPA 625/6-73/002.
11. Goldman, Charles R. & Horne, Alexander J. Limnology (1983) McGraw-Hill ISBN 0-07-023651-8 pp.87-88
12. United States Geological Survey (USGS), Denver, CO (2009). "Definitions of Quality-Assurance Data." Prepared by USGS Branch of Quality Systems, Office of Water Quality.
13. International Water Management Institute, Colombo, Sri Lanka (2010). "Helping restore the quality of drinking water after the tsunami." Success Stories. Issue 7. doi:10.5337/2011.0030
14. World Health Organization (2011). "WHO technical notes for emergencies." Water Engineering Development Centre, Loughborough University, Leicestershire, UK.
15. State of California Environmental Protection Agency Representative Sampling of Ground Water for Hazardous Substances (1994) pp.23-24
16. For an overview of the U.S. federal biomonitoring publications, see U.S. EPA, "Whole Effluent Toxicity Methods."

17. Methods for Measuring the Acute Toxicity of Effluents and Receiving Waters to Freshwater and Marine Organisms (Report). EPA. October 2002. EPA-821-R-02-012.
18. IOWATER (Iowa Department of Natural Resources). Iowa City, IA (2005). "Benthic Macroinvertebrate Key."
19. "Center for Coastal Monitoring and Assessment: Mussel Watch Contaminant Monitoring". Ccma.nos.noaa.gov. 2014-01-14. Retrieved 2015-09-04.
20. Dickens CWS and Graham PM. 2002. The Southern Africa Scoring System (SASS) version 5 rapid bioassessment for rivers "African Journal of Aquatic Science", 27:1-10.
21. "Guidelines for drinking-water quality, fourth edition". World Health Organization. Retrieved 2 April 2013.
22. International Organization for Standardization (ISO). "13.060: Water quality". Geneva, Switzerland. Retrieved 2011-07-04.
23. International Organization for Standardization (ISO). "91.140.60: Water supply systems". Retrieved 2011-07-04.
24. "Internet Scientific Publications".
25. Republic of South Africa, Department of Water Affairs, Pretoria (1996). "Water quality guidelines for South Africa: First Edition 1996."
26. Hodgson K, Manus L. A drinking water quality framework for South Africa. Water SA. 2006;32(5):673-678 .
27. National Archives, London, UK. "The Water Supply (Water Quality) Regulations 2000." 2000 No. 3184. 2000-12-08.
28. U.S. Clean Water Act, Section 303, 33 U.S.C. § 1313.
29. U.S. Clean Water Act, Section 303(d), 33 U.S.C. § 1313; Section 305(b), 33 U.S.C. § 1315(b).
30. "Program Overview: 303(d) Listing". Impaired Waters and TMDLs. EPA. 2016-10-24.
31. "National Water Quality Inventory Report to Congress". Water Data and Tools. EPA. 2016-08-18.
32. More information about water quality in the United States is available on EPA's "Surf Your Watershed" website.
33. "EPA Environmental Indicators Gateway,". USEPA. Retrieved 2010-06-16.
34. "Environmental Terminology and Discovery Service (ETDS)" (Information-base). European Environment Agency. Retrieved 2010-06-17.
35. Smeets, Edith; Rob Weterings (1999). "Environmental indicators:Typology and overview" (PDF). European Environment Agency. Retrieved 2009-03-16.
36. "State of the Great Lakes 2007 Highlights" (PDF). Environment Canada and the U.S. Environmental Protection Agency. 2007. Retrieved 2009-03-13.

37. "Environmental Indicators Gateway US EPA". U.S. Environmental Protection Agency. 2009. Retrieved 2009-03-16.
38. Canadian Environmental Sustainability Indicators
39. "UNEP-GEO". United Nations Environment Programme. 2009. Retrieved 2009-03-16.
40. Brown, L. R. (2011). World on the Edge. Earth Policy Institute. Norton. ISBN 978-0-393-08029-2.
41. Ditor, M.; O'Farrell, D.; Bond, W.; Engeland, J. (2001). "Guidelines for the development of sustainability indicators". Environment Canada and Canada Mortgage and Housing Corporation.
42. Rice, J.C.; Rochet, M-J. (2005). "A framework for selecting a suite of indicators for fisheries management". ICES Journal of Marine Science. 62 (3): 516–527. doi:10.1016/j.icesjms.2005.01.003.
43. Franson, Mary Ann Standard Methods for the Examination of Water and Wastewater 14th edition (1975) APHA, AWWA & WPCF ISBN 0-87553-078-8 pp.125-126
44. Franson, Mary Ann Standard Methods for the Examination of Water and Wastewater 14th edition (1975) APHA, AWWA & WPCF ISBN 0-87553-078-8 pp.89-98
45. Franson, Mary Ann Standard Methods for the Examination of Water and Wastewater 14th edition (1975) APHA, AWWA & WPCF ISBN 0-87553-078-8 pp.131-137
46. Franson, Mary Ann Standard Methods for the Examination of Water and Wastewater 14th edition (1975) APHA, AWWA & WPCF ISBN 0-87553-078-8 pp.99-100
47. Franson, Mary Ann Standard Methods for the Examination of Water and Wastewater 14th edition (1975) APHA, AWWA & WPCF ISBN 0-87553-078-8 pp.406-407
48. Goldman, Charles R. & Horne, Alexander J. Limnology (1983) McGraw-Hill ISBN 0-07-023651-8 p.111
49. Franson, Mary Ann Standard Methods for the Examination of Water and Wastewater 14th edition (1975) APHA, AWWA & WPCF ISBN 0-87553-078-8 pp.406-407
50. MacCutheon et al., 1983
51. US EPA (1984
52. Nduka et al., 2008
53. WHO (1984)
54. Franson, Mary Ann Standard Methods for the Examination of Water and Wastewater 14th edition (1975) APHA, AWWA & WPCF ISBN 0-87553-078-8 pp.309-315

55. Franson, Mary Ann Standard Methods for the Examination of Water and Wastewater 14th edition (1975) APHA, AWWA & WPCF ISBN 0-87553-078-8 pp.685-689

56. Franson, Mary Ann Standard Methods for the Examination of Water and Wastewater 14th edition (1975) APHA, AWWA & WPCF ISBN 0-87553-078-8 pp.875-877

57. Kashiwaya, Mamoru; Yoshimoto, Kameo (May 1980). "Tannery Wastewater Treatment by the Oxygen Activated Sludge Process". Journal (Water Pollution Control Federation). Alexandria, VA: Water Environment Federation. 52 (5): 999–1007. JSTOR 25040825.

58. U.S. Environmental Protection Agency (EPA), Washington, D.C. "Effluent Limitations Guidelines and Standards for the Steam Electric Power Generating Point Source Category." Final Rule. 2015-09-30.

59. European Environment Agency. Copenhagen, Denmark. "Indicator: Biochemical oxygen demand in rivers (2001)."

60. EPA (2002). "Section 7: Wastewater Characterization." Development Document for Final Effluent Limitations Guidelines and Standards for the Iron and Steel Manufacturing Point Source Category. Document no. EPA-821-R-02-004. pp. 7-1ff.

61. EPA (2000). "Permit Guidance Document: Pulp, Paper and Paperboard Manufacturing Point Source Category." Document no. EPA-821-B-00-003. pp. 4-1ff.

62. Tchobanoglous, G., Burton, F.L., and Stensel, H.D. (2003). Wastewater Engineering (Treatment Disposal Reuse) / Metcalf & Eddy, Inc. (4th ed.). McGraw-Hill Book Company. ISBN 0-07-041878-0.

63. Beychok, Milton R. (1967). Aqueous Wastes from Petroleum and Petrochemical Plants (1st ed.). John Wiley & Sons. LCCN 67019834.

64. Hobson, Tom (May 2004). "The Scoop on Oil Skimmers". Environmental Protection (formerly Water and Wastewater News). Dallas, TX: 1105 Media, Inc.

65. American Petroleum Institute (API) (February 1990). Management of Water Discharges: Design and Operations of Oil-Water Separators (1st ed.). American Petroleum Institute.

66. Beychok, Milton R. (December 1971). "Wastewater treatment". Hydrocarbon Processing: 109–112. ISSN 0887-0284.

67. Cleanawater (Virtus Industries Pty. Ltd.), Melbourne, Australia (2013). "Oil Water Separators Product Range." Brochure.

68. Primer for Municipal Waste water Treatment Systems (Report). Washington, DC: US Environmental Protection Agency. 2004. EPA 832-R-04-001..

69. Metcalf & Eddy, Inc. (1972). Wastewater Engineering. McGraw-Hill. ISBN 0-07-041675-3.

70. Cicek, V. (2013). "Corrosion and corrosion prevention in boilers". Cathodic protection: industrial solutions for protecting against corrosion. Hoboken, New Jersey: John Wiley & Sons. ISBN 9781118737880.

71. Huisman, L.; Wood, W.E. (1974). "Chapter 2. Filtration of Water Supplies". Slow Sand Filtration (PDF). Geneva: World Health Organization. ISBN 92-4-154037-0.

72. Buchan, James (2003). Crowded with genius: the Scottish enlightenment: Edinburgh's moment of the mind. New York: HarperCollins. ISBN 9780060558888.

73. Frerichs, Ralph R. "History of the Chelsea Waterworks". John Snow. Fielding School of Public Health, University of California, Los Angeles. Retrieved 2016-07-09.

74. Christman, Keith (September 1998). "The history of chlorine". WaterWorld. Tulsa, OK: PennWell. 14 (8): 66–67.

75. "Safe Water System" (PDF). Fact Sheet, World Water Forum 4 Update. Atlanta: US Centers for Disease Control and Prevention. June 2006.

76. "Household Water Treatment Guide". Centre for Affordable Water and Sanitation Technology, Canada. March 2008.

77. "Sand as a low-cost support for titanium dioxide photocatalysts". Materials Views. Wiley VCH.

78. Lindsten, Don C. (September 1984). "Technology transfer: Water purification, U.S. Army to the civilian community". The Journal of Technology Transfer. 9 (1): 57–59. doi:10.1007/BF02189057.

CHAPTER 7

Surface Runoff And Routing

Chapter Contents

- **Introduction**
- **Generation**
- **Routing**
- **Hydrograph**
- **Raster Hydrograph**
- **Shallow Water Equations**

Surface runoff refers to any water that flows over the surface of the Earth and into lakes, rivers and other bodies of water. This is in contrast to the water that is absorbed by soil or plants. Surface runoff is a crucial part of the water system. Water evaporates from lakes and rivers, but it must be replenished. While some rain or snow will fall directly into bodies of water, they also need some of the water that falls on land in order to maintain their size or flow. Surface runoff does carry the potential of sweeping up pesticides and other waste products and then polluting a small body of water.

INTRODUCTION

Surface runoff (also known as overland flow) is the flow of water that occurs when excess stormwater, meltwater, or other sources flows over the Earth's surface. This might occur because soil is saturated to full capacity, because rain arrives more quickly than soil can absorb it, or because impervious areas (roofs and pavement) send their runoff to surrounding soil that cannot absorb all of it. Surface runoff is a major component of the water cycle. It is the primary agent in soil erosion by water.

Runoff that occurs on the ground surface before reaching a channel is also called a nonpoint source. If a nonpoint source contains man-made contaminants, or natural forms of pollution (such as rotting leaves) the runoff is called nonpoint source pollution. A land area which produces runoff that drains to a common point is called a drainage basin. When runoff flows along the ground, it can pick up soil contaminants including petroleum, pesticides, or fertilizers that become discharge or nonpoint source pollution. In addition to causing water erosion and pollution, surface runoff in urban areas is a primary cause of urban flooding which can result in property damage, damp and mold in basements, and street flooding.

Runoff flowing into a stormwater drain

GENERATION

Surface runoff can be generated either by rainfall, snowfall or by the melting of snow, or glaciers.

Snow and glacier melt occur only in areas cold enough for these to form permanently. Typically snowmelt will peak in the spring and glacier melt in the summer, leading to pronounced flow maxima in rivers affected by them. The determining factor of the rate of melting of snow or glaciers is both air temperature and the duration of sunlight. In high mountain regions, streams frequently rise on sunny days and fall on cloudy ones for this reason.

Surface runoff from a hillside after soil is saturated

In areas where there is no snow, runoff will come from rainfall. However, not all rainfall will produce runoff because storage from soils can absorb light showers. On the extremely ancient soils of Australia and Southern Africa, proteoid roots with their extremely dense networks of root hairs can absorb so much rainwater as to prevent runoff even when substantial amounts of rain fall. In these regions, even on less infertile cracking clay soils, high amounts of rainfall and potential evaporation are needed to generate any surface runoff, leading to specialized adaptations to extremely variable (usually ephemeral) streams.

Infiltration Excess Overland Flow

This occurs when the rate of rainfall on a surface exceeds the rate at which water can infiltrate the ground, and any depression storage has already been filled. This is called flooding excess overland flow, Hortonian overland flow (after Robert E. Horton), or unsaturated overland flow. This more commonly occurs in arid and semi-arid regions, where rainfall intensities are high and the soil infiltration capacity is reduced because of surface sealing, or in paved areas. This occurs largely in city areas where pavements prevent water from flooding.

Saturation Excess Overland Flow

When the soil is saturated and the depression storage filled, and rain continues to fall, the rainfall will immediately produce surface runoff. The level of antecedent soil moisture is one factor affecting the time until soil becomes saturated. This runoff is called saturation excess overland flow or saturated overland flow.

Antecedent Soil Moisture

Soil retains a degree of moisture after a rainfall. This residual water moisture affects the soil's infiltration capacity. During the next rainfall event, the infiltration capacity will cause the soil to be saturated at a different rate. The higher the level of antecedent soil moisture, the more quickly the soil becomes saturated. Once the soil is saturated, runoff occurs.

Subsurface Return Flow

After water infiltrates the soil on an up-slope portion of a hill, the water may flow laterally through the soil, and exfiltrate (flow out of the soil) closer to a channel. This is called subsurface return flow or throughflow.

As it flows, the amount of runoff may be reduced in a number of possible ways: a small portion of it may evapotranspire; water may become temporarily stored in microtopographic depressions; and a portion of it may infiltrate as it flows overland. Any remaining surface water eventually flows into a receiving water body such as a river, lake, estuary or ocean.

Human influence

Urban surface water runoff

Urbanization increases surface runoff by creating more impervious surfaces such as pavement and buildings that do not allow percolation of the water down through the soil to the aquifer. It is instead forced directly into streams or storm water runoff drains, where erosion and siltation can be major problems, even when flooding is not. Increased runoff reduces groundwater recharge, thus lowering the water table and making droughts worse, especially for farmers and others who depend on the water wells.

When anthropogenic contaminants are dissolved or suspended in runoff, the human impact is expanded to create water pollution. This pollutant load can reach various receiving waters such as streams, rivers, lakes, estuaries and oceans with resultant water chemistry changes to these water systems and their related ecosystems.

A 2008 report by the United States National Research Council identified urban stormwater as a leading source of water quality problems in the U.S.

As humans continue to alter the climate through the addition of greenhouse gases to the atmosphere, precipitation patterns are expected to change as the atmospheric capacity for water vapor increases. This will have direct consequences on runoff amounts.

Effects of Surface Runoff

Erosion and Deposition

Surface runoff can cause erosion of the Earth's surface; eroded material may be deposited a considerable distance away. There are four main types of soil erosion by water: splash erosion, sheet erosion, rill erosion and gully erosion. Splash erosion is the result of mechanical collision of raindrops with the soil surface: soil particles which are dislodged by the impact then move with the surface runoff. Sheet erosion is the overland transport of sediment by runoff without a well defined channel. Soil surface roughness causes may cause runoff to become concentrated into narrower flow paths: as these incise, the small but well-defined channels which are formed are known as rills. These channels can be as small as one centimeter wide or as large as several meters. If runoff continue to incise and enlarge rills, they may eventually grow to become gullies. Gully erosion can transport large amounts of eroded material in a small time period.

Reduced crop productivity usually results from erosion, and these effects are studied in the field of soil conservation. The soil particles carried in runoff vary in size from about .001 millimeter to 1.0 millimeter in diameter. Larger particles settle over short transport distances, whereas small particles can be carried over long distances suspended in the water column. Erosion of silty soils that contain smaller particles generates turbidity and diminishes light transmission, which disrupts aquatic ecosystems.

Entire sections of countries have been rendered unproductive by erosion. On the high central plateau of Madagascar, approximately ten percent of that country's land area, virtually the entire landscape is devoid of vegetation, with erosive gully furrows typically in excess of 50 meters deep and one kilometer wide. Shifting cultivation is a farming system which sometimes incorporates the slash and burn method in some regions of the world. Erosion causes loss of the fertile top soil and reduces its fertility and quality of the agricultural produce.

Modern industrial farming is another major cause of erosion. In some areas in the American corn belt, more than 50 percent of the original topsoil has been carried away within the last 100 years.

Soil erosion by water on intensively-tilled farmland.

Willow hedge strengthened with fascines for the limitation of runoff, north of France.

Environmental Effects

The principal environmental issues associated with runoff are the impacts to surface water, groundwater and soil through transport of water pollutants to these systems. Ultimately these consequences translate into human health risk, ecosystem disturbance and aesthetic impact to water resources. Some of the contaminants that create the greatest impact to surface waters arising from runoff are petroleum substances, herbicides and fertilizers. Quantitative uptake by surface runoff of pesticides and other contaminants has been studied since the 1960s, and early on contact of pesticides with water was known to enhance phytotoxicity. In the case of surface waters, the impacts translate to water pollution, since the streams and rivers have received runoff carrying various chemicals or sediments. When surface waters are used as potable water supplies, they can be compromised regarding health risks and drinking water aesthetics (that is, odor, color and turbidity effects). Contaminated surface waters risk altering the metabolic processes of the aquatic species that they host; these alterations can lead to death, such as fish kills, or alter the balance of populations present. Other specific impacts are on animal mating, spawning, egg and larvae viability, juvenile survival and plant productivity. Some researches show surface runoff of pesticides, such as DDT, can alter the gender of fish species genetically, which transforms male into female fish.

Surface runoff occurring within forests can supply lakes with high loads of mineral nitrogen and phosphorus leading to eutrophication. Runoff waters within coniferous forests are also enriched with humic acids and can lead to humification of water bodies

In the case of groundwater, the main issue is contamination of drinking water, if the aquifer is abstracted for human use. Regarding soil contamination, runoff waters can have two important pathways of concern. Firstly, runoff water can extract soil contaminants and carry them in the form of water pollution to even more sensitive aquatic habitats. Secondly, runoff can deposit contaminants on pristine soils, creating health or ecological consequences.

Agricultural Issues

The other context of agricultural issues involves the transport of agricultural chemicals (nitrates, phosphates, pesticides, herbicides etc.) via surface runoff. This result occurs when chemical use is excessive or poorly timed with respect to high precipitation. The resulting contaminated runoff represents not only a waste of agricultural chemicals, but also an environmental threat to downstream ecosystems.

Flooding

Flooding occurs when a watercourse is unable to convey the quantity of runoff flowing downstream. The frequency with which this occurs is described by a return period. Flooding is a natural process, which maintains ecosystem composition and processes, but it can also be altered by land use changes such as river engineering. Floods can be both beneficial to societies or cause damage. Agriculture along the Nile floodplain took advantage of the seasonal flooding that deposited nutrients beneficial for crops. However, as the number and susceptibility of settlements increase, flooding increasingly becomes a natural hazard. In urban areas, surface runoff is the primary cause of urban flooding, known for its repetitive and costly impact on communities. Adverse impacts span loss of life, property damage, contamination of water supplies, loss of crops, and social dislocation and temporary homelessness. Floods are among the most devastating of natural disasters. Mitigation and Treatment

Mitigation of adverse impacts of runoff can take several forms:

Land use development controls aimed at minimizing impervious surfaces in urban areas

- Erosion controls for farms and construction sites
- Flood control and retrofit programs, such as green infrastructure
- Chemical use and handling controls in agriculture, landscape maintenance, industrial use, etc.

Runoff holding ponds (Uplands neighborhood of North Bend, Washington)

Land use controls. Many world regulatory agencies have encouraged research on methods of minimizing total surface runoff by avoiding unnecessary hardscape. Many municipalities have produced guidelines and codes (zoning and related ordinances) for land developers that encourage minimum width sidewalks, use of pavers set in earth for driveways and walkways and other design techniques to allow maximum water infiltration in urban settings. An example land use control program can be seen in the city of Santa Monica, California.

Erosion controls have appeared since medieval times when farmers realized the importance of contour farming to protect soil resources. Beginning in the 1950s these agricultural methods became increasingly more sophisticated. In the 1960s some state and local governments began to focus their efforts on mitigation of construction runoff by requiring builders to implement erosion and sediment controls (ESCs). This included such techniques as: use of straw bales and barriers to slow runoff on slopes, installation of silt fences, programming construction for months that have less rainfall and minimizing extent and duration of exposed graded areas. Montgomery County, Maryland implemented the first local government sediment control program in 1965, and this was followed by a statewide program in Maryland in 1970.

Flood control programs as early as the first half of the twentieth century became quantitative in predicting peak flows of riverine systems. Progressively strategies have been developed to minimize peak flows and also to reduce channel velocities. Some of the techniques commonly applied are: provision of holding ponds (also called detention basins) to buffer riverine peak flows, use of energy dissipators in channels to reduce stream velocity and land use controls to minimize runoff.

Chemical use and handling. Following enactment of the U.S. Resource Conservation and Recovery Act (RCRA) in 1976, and later the Water Quality Act of 1987, states and cities have become more vigilant in controlling the containment and storage of toxic chemicals, thus preventing releases and leakage. Methods commonly applied are: requirements for double containment of underground storage tanks, registration of hazardous materials usage, reduction in numbers of allowed pesticides and more stringent regulation of fertilizers and herbicides in landscape maintenance. In many industrial cases, pretreatment of wastes is required, to minimize escape of pollutants into sanitary or stormwater sewers.

The U.S. Clean Water Act (CWA) requires that local governments in urbanized areas (as defined by the Census Bureau) obtain stormwater discharge permits for their drainage systems. Essentially this means that the locality must operate a stormwater management program for all surface runoff that enters the municipal separate storm sewer system ("MS4"). EPA and state regulations and related publications outline six basic components that each local program must contain:

- Public education (informing individuals, households, businesses about ways to avoid stormwater pollution)
- Public involvement (support public participation in implementation of local programs)
- Illicit discharge detection & elimination (removing sanitary sewer or other non-stormwater connections to the MS4)
- Construction site runoff controls (i.e. erosion & sediment controls)
- Post-construction (i.e. permanent) stormwater management controls
- Pollution prevention and "good housekeeping" measures (e.g. system maintenance).

Other property owners which operate storm drain systems similar to municipalities, such as state highway systems, universities, military bases and prisons, are also subject to the MS4 permit requirements.

Measurement and Mathematical Modeling

Runoff is analyzed by using mathematical models in combination with various water quality sampling methods. Measurements can be made using continuous automated water quality analysis instruments targeted on pollutants such as specific organic or inorganic chemicals, pH, turbidity etc. or targeted on secondary indicators such as dissolved oxygen.

Measurements can also be made in batch form by extracting a single water sample and conducting any number of chemical or physical tests on that sample.

In the 1950s or earlier hydrology transport models appeared to calculate quantities of runoff, primarily for flood forecasting. Beginning in the early 1970s computer models were developed to analyze the transport of runoff carrying water pollutants, which considered dissolution rates of various chemicals, infiltration into soils and ultimate pollutant load delivered to receiving waters. One of the earliest models addressing chemical dissolution in runoff and resulting transport was developed in the early 1970s under contract to the United States Environmental Protection Agency (EPA). This computer model formed the basis of much of the mitigation study that led to strategies for land use and chemical handling controls.

Other computer models have been developed (such as the DSSAM Model) that allow surface runoff to be tracked through a river course as reactive water pollutants. In this case the surface runoff may be considered to be a line source of water pollution to the receiving waters.

ROUTING

In hydrology, routing is a technique used to predict the changes in shape of a hydrograph as water moves through a river channel or a reservoir. In flood forecasting, hydrologists may want to know how a short burst of intense rain in an area upstream of a city will change as it reaches the city. Routing can be used to determine whether the pulse of rain reaches the city as a deluge or a trickle.

Routing also can be used to predict the hydrograph shape (and thus lowland flooding potential) subsequent to multiple rainfall events in different sub-catchments of the watershed. Timing and duration of the rainfall events, as well as factors such as antecedent moisture conditions, overall watershed shape, along with subcatchment-area shapes, land slopes (topography/physiography), geology/hydrogeology (i.e. forests and aquifers can serve as giant sponges that absorb rainfall and slowly release it over subsequent weeks and months), and stream-reach lengths all play a role here. The result can be an additive effect (i.e. a large flood if each subcatchment's respective hydrograph peak arrives at the watershed mouth at the same point in time, thereby effectively causing a "stacking" of the hydrograph peaks), or a more distributed-in-time effect (i.e. a lengthy but relatively modest flood, effectively attenuated in time, as the individual subcatchment peaks arrive at the mouth of the main watershed channel in orderly succession).

Other uses of routing include reservoir and channel design, floodplain studies and watershed simulations.

If the water flow at a particular point, A, in a stream is measured over time with a flow gauge, this information can be used to create a hydrograph. A short period of intense rain, normally called a flood event, can cause a bulge in the graph, as the increased water travels down the river, reaches the flow gauge at A, and passes along it.

If another flow gauge at B, downstream of A is set up, one would expect the graph's bulge (or floodwave) to have the same shape. However, the shape of the river and flow resistance within a river (from the river bed, for example) can affect the shape of the floodwave. Oftentimes, the floodwave will be attenuated (have a reduced peak flow).

Routing techniques can be broadly classified as follows:

Hydraulic (Or Distributed) Routing

Hydraulic routing is based on the solution of partial differential equations of unsteady open-channel flow. The equations used are the St. Venant equations or the dynamic wave equations.

The hydraulic models (e.g. ***dynamic*** and ***diffusion*** wave models) require the gathering of a lot of data related to river geometry and morphology and consume a lot of computer resources in order to solve the Saint-Venant equations numerically.

Hydrologic (Or Lumped) Routing

Hydrologic routing uses the continuity equation for hydrology. In its simplest form, inflow to the river reach is equal to the outflow of the river reach plus the change of storage:

$$I = O + \frac{\Delta S}{\Delta t}$$

, where

- I is average inflow to the reach during Δt
- O is average outflow from the reach during Δt ; and
- S is the water currently in the reach (known as storage)

The hydrologic models (e.g. linear and nonlinear Muskingum models) need to estimate hydrologic parameters using recorded data in both upstream and downstream sections of rivers and/or by applying robust optimization techniques to solve the one-dimensional conservation of mass and storage-continuity equation.

Semi-distributed routing

Nowadays, semi-distributed models such as Muskingum–Cunge family procedures are also available. Simple physical concepts and common river characteristics such as channel geometry, reach length, roughness coefficient, and slope are used to estimate the model parameters without complex and expensive numerical solutions.

Selection of routing procedure

In general, based on the available field data and goals of the project, one of routing procedures is selected.

HYDROGRAPH

A hydrograph is a graph showing the rate of flow (discharge) versus time past a specific point in a river, or other channel or conduit carrying flow. The rate of flow is typically expressed in cubic meters or cubic feet per second (cms or cfs).

It can also refer to a graph showing the volume of water reaching a particular outfall, or location in a sewerage network. Graphs are commonly used in the design of sewerage, more specifically, the design of surface water sewerage systems and combined sewers.

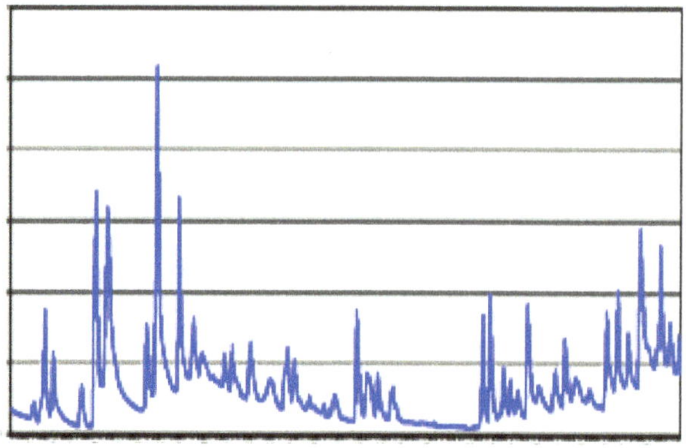

Stream hydrograph. Increases in stream flow follow rainfall or snowmelt. The gradual decay in flow after the peaks reflects diminishing supply from groundwater.

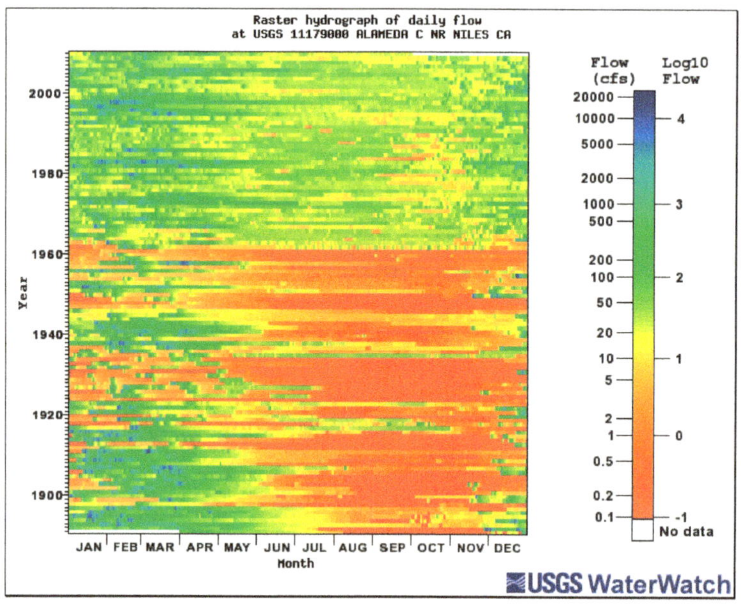

Raster hydrograph. The entire flow record and patterns representing different timescales can be visualized.

Terminology

The discharge is measured at a specific point in a river and is typically time variant.

- Rising limb: The rising limb of hydro graph, also known as concentration curve, reflects a prolonged increase in discharge from a catchment area, typically in response to a rainfall event
- Recession (or falling) limb: The recession limb extends from the peak flow rate

onward. The end of stormflow (aka quickflow or direct runoff) and the return to groundwater-derived flow (base flow) is often taken as the point of inflection of the recession limb. The recession limb represents the withdrawal of water from the storage built up in the basin during the earlier phases of the hydrograph.

- Peak discharge: the highest point on the hydro graph when the rate of discharge is greatest
- Lag time: the time interval from the center of mass of rainfall excess to the peak of the resulting hydrograph
- Time to peak: time interval from the start of the resulting hydro graph
- Discharge: the rate of flow (volume per unit time) passing a specific location in a river or other channel

Types of hydrograph can include:

- Storm hydrographs
- Flood hydrographs
- Annual hydrographs aka regimes
- Direct Runoff Hydrograph
- Effective Runoff Hydrograph
- Raster Hydrograph
- Storage opportunities in the drainage network (e.g., lakes, reservoirs, wetlands, channel and bank storage capacity)

Baseflow Separation

A stream hydrograph is commonly conceptualized to include a baseflow component' and a runoff component. The former represents the relatively steady contribution to stream discharge from groundwater return flow, while the latter represents the additional streamflow contributed by subsurface flow and surface flow/runoff]].

The separation of baseflow from direct runoff in a hydrograph is often of interest to hydrologists, planners, and engineers, as it aids in determining the influence of different hydrologic processes on discharge from the subject catchment. Because the timing, magnitude, and duration of groundwater return flow differs so greatly from that of direct runoff, separating and understanding the influence of these distinct processes is key to analyzing and simulating the likely hydrologic effects of various land use, water use, weather, and climate conditions and changes.

However, the process of separating "baseflow" from "direct runoff" is an inexact science. In part this is because these two concepts are not, themselves, entirely distinct and unrelated. Return flow from groundwater increases along with overland flow from saturated or impermeable areas during and after a storm event; moreover, a particular water molecule can easily move through both pathways en route to the watershed outlet. Therefore, separation of a purely "baseflow component" in a hydrograph is a somewhat arbitrary exercise. Nevertheless, various graphical and empirical techniques have been developed to perform these hydrograph separations. The separation of base flow from direct runoff can be an important first step in developing rainfall-runoff models for a watershed of interest—for example, in developing and applying unit hydrographs as described below.

Unit hydrograph

A unit hydrograph (UH) is the hypothetical unit response of a watershed (in terms of runoff volume and timing) to a unit input of rainfall. It can be defined as the direct runoff hydrograph (DRH) resulting from one unit (e.g., one cm or one inch) of effective rainfall occurring uniformly over that watershed at a uniform rate over a unit period of time. As a UH is applicable only to the direct runoff component of a hydrograph (i.e., surface runoff), a separate determination of the baseflow component is required.

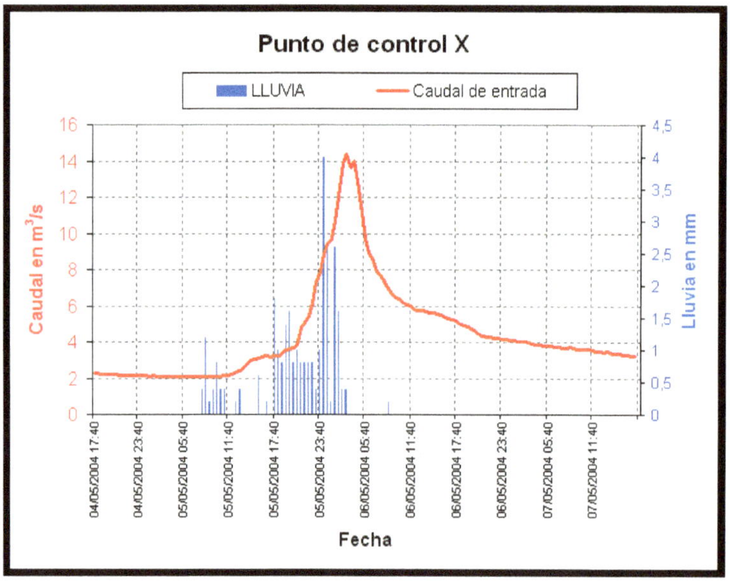

A UH is specific to a particular watershed, and specific to a particular length of time corresponding to the duration of the effective rainfall. That is, the UH is specified as being the 1-hour, 6-hour, or 24-hour UH, or any other length of time up to the time of concentration of direct runoff at the watershed outlet. Thus, for a given watershed, there can be many unit hydrographs, each one corresponding to a different duration of effective rainfall.

The UH technique provides a practical and relatively easy-to-apply tool for quantifying the effect of a unit of rainfall on the corresponding runoff from a particular drainage basin. UH theory assumes that a watershed's runoff response is linear and time-invariant, and that the effective rainfall occurs uniformly over the watershed. In the real world, none of these assumptions are strictly true. Nevertheless, application of UH methods typically yields a reasonable approximation of the flood response of natural watersheds. The linear assumptions underlying UH theory allows for the variation in storm intensity over time (i.e., the storm hyetograph) to be simulated by applying the principles of superposition and proportionality to separate storm components to determine the resulting cumulative hydrograph. This allows for a relatively straightforward calculation of the hydrograph response to any arbitrary rain event.

An instantaneous unit hydrograph is a further refinement of the concept; for an IUH, the input rainfall is assumed to all take place at a discrete point in time (obviously, this isn't the case for actual rainstorms). Making this assumption can greatly simplify the analysis involved in constructing a unit hydrograph, and it is necessary for the creation of a geomorphologic instantaneous unit hydrograph.

The creation of a GIUH is possible given nothing more than topologic data for a particular drainage basin. In fact, only the number of streams of a given order, the mean length of streams of a given order, and the mean land area draining directly to streams of a given order are absolutely required (and can be estimated rather than explicitly calculated if necessary). It is therefore possible to calculate a GIUH for a basin without any data about stream height or flow, which may not always be available.

Subsurface Hydrology Hydrograph

In subsurface hydrology (hydrogeology), a hydrograph is a record of the water level (the observed hydraulic head in wells screened across an aquifer).

Typically, a hydrograph is recorded for monitoring of heads in aquifers during non-test conditions (e.g., to observe the seasonal fluctuations in an aquifer). When an aquifer test is being performed, the resulting observations are typically called drawdown, since they are subtracted from pre-test levels and often only the change in water level is dealt with.

RASTER HYDROGRAPH

Raster hydrographs are pixel-based plots for visualizing and identifying variations and changes in large multidimensional data sets. Originally developed by Keim (2000) they were first applied in hydrology by Koehler (2004) as a means of highlighting inter-annual and intra-annual changes in streamflow. The raster hydrographs in WaterWatch, like those developed by Koehler, depict years on the y-axis and days along the x-axis. Users can choose to plot streamflow (actual values or log values), streamflow percentile, or streamflow class (from 1, for low flow, to 7 for high flow), for Daily, 7-Day, 14-Day, and 28-Day streamflow. For a more comprehensive description of raster hydrographs.

SHALLOW WATER EQUATIONS

The shallow water equations (also called Saint-Venant equations in its unidimensional form, after Adhémar Jean Claude Barré de Saint-Venant) are a set of hyperbolic partial differential equations (or parabolic if viscous shear is considered) that describe the flow below a pressure surface in a fluid (sometimes, but not necessarily, a free surface). The shallow water equations can also be simplified to the commonly used 1-D Saint-Venant equation.

The equations are derived from depth-integrating the Navier–Stokes equations, in the case where the horizontal length scale is much greater than the vertical length scale. Under this condition, conservation of mass implies that the vertical velocity of the fluid is small. It can be shown from the momentum equation that vertical pressure gradients are nearly hydrostatic, and that horizontal pressure gradients are due to the displacement of the pressure surface, implying that the horizontal velocity field is constant throughout the depth of the fluid. Vertically integrating allows the vertical velocity to be removed from the equations. The shallow water equations are thus derived.

While a vertical velocity term is not present in the shallow water equations, note that this velocity is not necessarily zero. This is an important distinction because, for example, the vertical velocity cannot be zero when the floor changes depth, and thus if it were zero only flat floors would be usable with the shallow water equations. Once a solution (i.e. the horizontal velocities and free surface displacement) has been found, the vertical velocity can be recovered via the continuity equation.

Situations in fluid dynamics where the horizontal length scale is much greater than the vertical length scale are common, so the shallow water equations are widely applicable. They are used with Coriolis forces in atmospheric and oceanic modeling, as a simplification of the primitive equations of atmospheric flow.

Shallow water equation models have only one vertical level, so they cannot directly encompass any factor that varies with height. However, in cases where the mean state is sufficiently simple, the vertical variations can be separated from the horizontal and several sets of shallow water equations can describe the state.

Equations

Conservative Form

The shallow water equations are derived from equations of conservation of mass and conservation of linear momentum (the Navier–Stokes equations), which hold even when the assumptions of shallow water break down, such as across a hydraulic jump. In the case of no Coriolis, frictional or viscous forces, the shallow-water equations are:

$$\frac{\partial \eta}{\partial t} + \frac{\partial(\eta u)}{\partial x} + \frac{\partial(\eta v)}{\partial y} = 0$$

$$\frac{\partial(\eta u)}{\partial t} + \frac{\partial}{\partial x}\left(\eta u^2 + \frac{1}{2}g\eta^2\right) + \frac{\partial(\eta uv)}{\partial y} = 0$$

$$\frac{\partial(\eta v)}{\partial t} + \frac{\partial(\eta uv)}{\partial x} + \frac{\partial}{\partial y}\left(\eta v^2 + \frac{1}{2}g\eta^2\right) = 0.$$

Here η is the total fluid column height, and the 2D vector (u,v) is the fluid's horizontal velocity, averaged across the vertical column. g is acceleration due to gravity. The first equation is derived from mass conservation, the second two from momentum conservation.

Non-Conservative Form

Expanding the derivatives in the above using the product rule, we obtain non-conservative forms of the shallow-water equations. Since velocities are not subject to a fundamental conservation equation, the non-conservative forms do not hold across a shock or hydraulic jump. When we include also the appropriate terms for Coriolis, frictional and viscous forces, we obtain:

$$\frac{\partial u}{\partial t}+u\frac{\partial u}{\partial x}+v\frac{\partial u}{\partial y}-fv=-g\frac{\partial h}{\partial x}-bu,$$

$$\frac{\partial v}{\partial t}+u\frac{\partial v}{\partial x}+v\frac{\partial v}{\partial y}+fu=-g\frac{\partial h}{\partial y}-bv,$$

$$\frac{\partial h}{\partial t}=-\frac{\partial}{\partial x}\Big(u\,(H+h)\Big)-\frac{\partial}{\partial y}\Big(v\,(H+h)\Big),$$

where

u	is the velocity in the x direction, or zonal velocity
v	is the velocity in the y direction, or meridional velocity
h	is the height deviation of the horizontal pressure surface from its mean height H
H	is the mean height of the horizontal pressure surface
g	is the acceleration due to gravity
f	is the Coriolis coefficient associated with the Coriolis force, on Earth equal to $2\Omega\sin(\varphi)$, where Ω is the angular rotation rate of the Earth ($\pi/12$ radians/hour), and φ is the latitude
b	is the viscous drag coefficient

It is often the case that the terms quadratic in u and v, which represent the effect of bulk advection, are small compared to the other terms. This is called geostrophic balance, and is equivalent to saying that the Rossby number is small. Assuming also that the wave height is very small compared to the mean height ($h \ll H$), we have:

$$\frac{\partial u}{\partial t}-fv=-g\frac{\partial h}{\partial x}-bu,$$

$$\frac{\partial v}{\partial t}+fu=-g\frac{\partial h}{\partial y}-bv,$$

$$\frac{\partial h}{\partial t}=-H\Big(\frac{\partial u}{\partial x}+\frac{\partial v}{\partial y}\Big)$$

One-dimensional Saint-Venant equations

The one-dimensional (1-D) Saint-Venant equations were derived by Adhémar Jean Claude Barré de Saint-Venant, and are commonly used to model transient open-channel flow and surface runoff. They can be viewed as a contraction of the two-dimensional (2-D) shallow water equations, which are also known as the two-dimensional Saint-Venant equations. The 1-D Saint-Venant equations contain to a certain extend the main characteristics of the channel cross-sectional shape.

The 1-D equations are used extensively in computer models such as HEC-RAS, SWMM5, ISIS, InfoWorks, Flood Modeller, SOBEK 1DFlow, MIKE 11, and MIKE SHE because they are significantly easier to solve than the full shallow water equations. Common applications of the 1-D Saint-Venant equations include flood routing along rivers (including evaluation of measures to reduce the risks of flooding), dam break analysis, storm pulses in an open channel, as well as storm runoff in overland flow.

The Equations

The system of partial differential equations which describe the 1-D incompressible flow in an open channel of arbitrary cross section – as derived and posed by Saint-Venant in his 1871 paper (equations 19 & 20) – are:

$$\frac{\partial A}{\partial t} + \frac{\partial (Au)}{\partial x} = 0 \qquad \text{and}$$

$$\frac{\partial u}{\partial t} + u\frac{\partial u}{\partial x} + g\frac{\partial \zeta}{\partial x} = -\frac{P}{A}\frac{\tau}{\rho},$$

where x is the space coordinate along the channel axis, t denotes time, A(x,t) is the cross-sectional area of the flow at location x, u(x,t) is the flow velocity, ζ(x,t) is the free surface elevation and τ(x,t) is the wall shear stress along the wetted perimeter P(x,t) of the cross section at x. Further ρ is the (constant) fluid density and g is the gravitational acceleration.

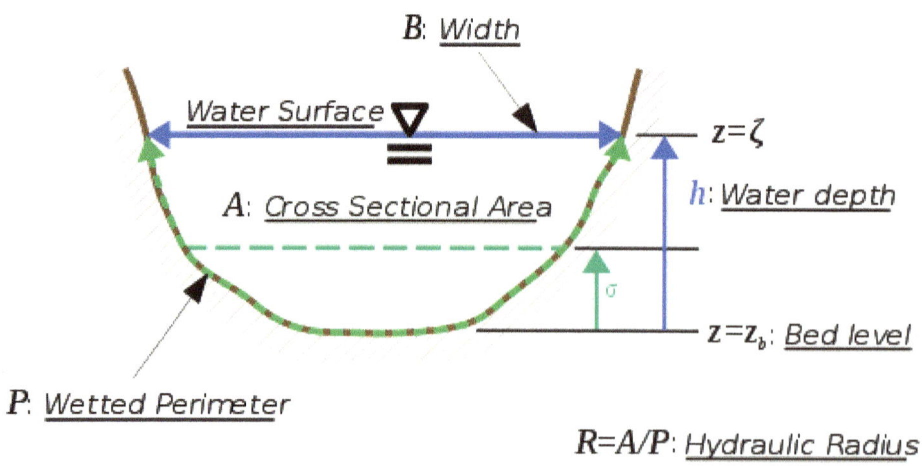

Cross section of the open channel.

Closure of the hyperbolic system of equations (**1**)–(**2**) is obtained from the geometry of cross sections – by providing a functional relationship between the cross-sectional area A and the surface elevation ζ at each position x. For example, for a rectangular cross section, with constant channel width B and channel bed elevation z_b, the cross sectional area is: $A = B(\zeta - z_b)$. The wall shear stress τ is dependent on the flow velocity u, they can be related by using e.g. the Darcy–Weisbach equation.

Further, equation (**1**) is the continuity equation, expressing conservation of water volume for this incompressible homogeneous fluid. Equation (**2**) is the momentum equation, giving the balance between forces and momentum change rates.

The instantaneous water depth is $h(x,t) = \zeta(x,t) - z_b(x)$, with $z_b(x)$ the bed level (i.e. elevation of the lowest point in the bed above datum. The bed slope S(x), friction slope $S_f(x,t)$ and hydraulic radius R(x,t) are defined as:

$$S = -\frac{\mathrm{d}z_b}{\mathrm{d}x}, \qquad S_f = \frac{\tau}{\rho g R} \qquad \text{and} \qquad R = \frac{A}{P}.$$

Consequently, the momentum equation (**2**) can be written as:

$$\frac{\partial u}{\partial t} + u\frac{\partial u}{\partial x} + g\frac{\partial h}{\partial x} + g\,(S_f - S) = 0.$$

Further note that – for non-moving channel walls – in the continuity equation (**1**): $\partial A/\partial t = B\,\partial\zeta/\partial t = B\,\partial h/\partial t$, with B(x,t) the momentary width of the free surface in the channel cross section at location x.

Equivalently, the momentum equation (**3**) can also be cast in the so-called conservation form, through some algebraic manipulations on the Saint-Venant equations, (**1**) and (**3**). In terms of the discharge $Q = Au$:

$$\frac{\partial Q}{\partial t} + \frac{\partial}{\partial x}\left(\frac{Q^2}{A} + g\,I_1\right) + g\,A\,(S_f - S) - g\,I_2 = 0,$$

where A, I_1 and I_2 are functions of the channel geometry, described in the terms of the channel width $B(\sigma,x)$. Here σ is the height above the lowest point in the cross section at location x, i.e. the height above the bed level $z_b(x)$ (the lowest point in the cross section):

$$A(\sigma,x) = \int_0^\sigma B(\sigma',x)\,\mathrm{d}\sigma',$$

$$I_1(\sigma,x) = \int_0^\sigma (\sigma-\sigma')\,B(\sigma',x)\,\mathrm{d}\sigma' \qquad \text{and}$$

$$I_2(\sigma,x) = \int_0^\sigma (\sigma-\sigma')\,\frac{\partial B(\sigma',x)}{\partial x}\,\mathrm{d}\sigma'.$$

Above – in the momentum equation (**4**) in conservation form – A, I_1 and I_2 are evaluated at $\sigma = h(x,t)$. The term g I_1 describes the hydrostatic force in a certain cross section. And, for a non-prismatic channel, g I_2 gives the effects of geometry variations along the channel axis x.

In applications, depending on the problem at hand, there often is a preference for using either the momentum equation in non-conservation form, (**2**) or (**3**), or the conservation form (**4**). For instance in case of the description of hydraulic jumps, the conservation form is preferred since the momentum flux is continuous across the jump.

Characteristics

The Saint-Venant equations (**1**)–(**2**) can be analyzed using the method of characteristics. The two celerities dx/dt on the characteristic curves are:

$$\frac{\mathrm{d}x}{\mathrm{d}t} = u \pm c, \qquad \text{with} \qquad c = \sqrt{\frac{gA}{B}}.$$

The Froude number $F = |u| / c$ determines whether the flow is subcritical ($F < 1$) or supercritical ($F > 1$).

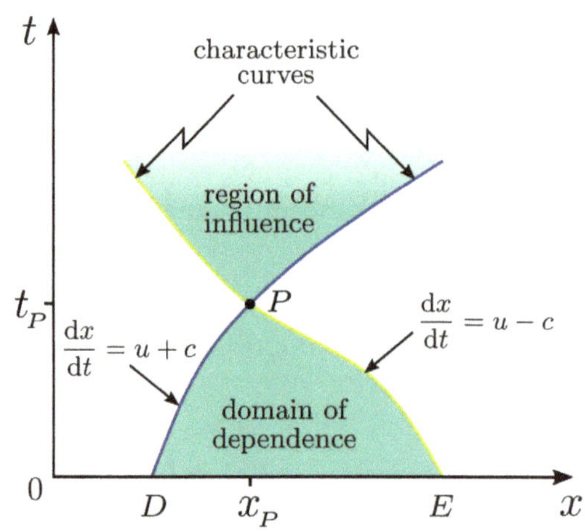

Characteristics, domain of dependence and region of influence, associated with location $P = (x_P, t_P)$ in space x and time t.

For a rectangular and prismatic channel of constant width B, i.e. with $A = B\,h$ and $c = \sqrt{gh}$, the Riemann invariants are:

$$r_+ = u + 2\sqrt{gh} \qquad \text{and}$$
$$r_- = u - 2\sqrt{gh},$$

so the equations in characteristic form are:

$$\frac{\mathrm{d}}{\mathrm{d}t}\left(u + 2\sqrt{gh}\right) = g\left(S - S_f\right) \qquad \text{along} \quad \frac{\mathrm{d}x}{\mathrm{d}t} = u + \sqrt{gh} \quad \text{and}$$
$$\frac{\mathrm{d}}{\mathrm{d}t}\left(u - 2\sqrt{gh}\right) = g\left(S - S_f\right) \qquad \text{along} \quad \frac{\mathrm{d}x}{\mathrm{d}t} = u - \sqrt{gh}.$$

The Riemann invariants and method of characteristics for a prismatic channel of arbitrary cross-section are described by Didenkulova & Pelinovsky (2011).

The characteristics and Riemann invariants provide important information on the behavior of the flow, as well as that they may be used in the process of obtaining (analytical or numerical) solutions.

Derived Modelling

Dynamic Wave

The dynamic wave is the term used to describe the full 1-dimensional Saint-Venant equation. It is numerically challenging to solve, but is valid for all channel flow scenarios. The dynamic wave is used for modeling transient storms in modeling programs including HEC-RAS, InfoWorks, MIKE 11, Wash 123d and SWMM.

Kinematic Wave

For the kinematic wave it is assumed that the flow is uniform, and that the friction slope is approximately equal slope of the channel. This simplifies the full Saint-Venant equation to the kinematic wave:

$$S_f - S = 0.$$

The kinematic wave is valid when the change in wave height over distance and velocity over distance and time is negligible relative to the bed slope, e.g. for shallow flows over steep slopes. The kinematic wave is used in HEC-HMS.

Diffusive Wave

For the diffusive wave it is assumed that the inertial terms are less than the gravity, friction, and pressure terms. The diffusive wave can therefore be more accurately described as a non-inertia wave, and is written as:

$$g\frac{\partial h}{\partial x} + g(S_f - S) = 0.$$

The diffusive wave is valid when the inertial acceleration is much smaller than all other forms of acceleration, or in other words when there is primarily subcritical flow. Models that use the diffusive wave assumption include MIKE SHE and LISFLOOD-FP.

Derivation from Navier–Stokes equations

The 1-D Saint-Venant momentum equation can be derived from the Navier–Stokes equations that describe fluid motion. The x-component of the Navier–Stokes equations – when expressed in Cartesian coordinates in the x-direction – can be written as:

$$g\frac{\partial h}{\partial x} + g(S_f - S) = 0.$$

where u is the velocity in the x-direction, v is the velocity in the y-direction, w is the velocity in the z-direction, t is time, p is the pressure, ρ is the density of water, v is the kinematic viscosity, and f_x is the body force in the x-direction.

1. If it is assumed that friction is taken into account as a body force, then v can be assumed as zero so:

$$\nu\left(\frac{\partial^2 u}{\partial x^2} + \frac{\partial^2 u}{\partial y^2} + \frac{\partial^2 u}{\partial z^2}\right) = 0.$$

2. Assuming one-dimensional flow in the *x*-direction it follows that:

$$v\frac{\partial u}{\partial y} + w\frac{\partial u}{\partial z} = 0$$

3. Assuming also that the pressure distribution is approximately hydrostatic it follows that:

$$p = \rho g h$$

   or in differential form:

$$\partial p = \rho g(\partial h).$$

   And when these assumptions are applied to the *x*-component of the Navier–Stokes equations:

$$-\frac{\partial p}{\partial x}\frac{1}{\rho} = -\frac{1}{\rho}\frac{\rho g\,(\partial h)}{\partial x} = -g\frac{\partial h}{\partial x}.$$

4. There are 2 body forces acting on the channel fluid, gravity, and friction:

$$f_x = f_{x,g} + f_{x,f}$$

   where $f_{x,g}$ is the body force due to gravity and $f_{x,f}$ is the body force due to friction.

5. $f_{x,g}$ can be calculated using basic physics and trigonometry:

$$F_g = (\sin\theta)gM$$

   where F_g is the force of gravity in the *x*-direction, *θ* is the angle, and *M* is the mass.

where Fg is the force of gravity in the x-direction, θ is the angle, and M is the mass. The expression for sin θ can be simplified using trigonometry as

$$\sin\theta = \frac{\text{opp}}{\text{hyp}}.$$

For small θ (reasonable for almost all streams) it can be assumed that:

$$\sin\theta = \tan\theta = \frac{\text{opp}}{\text{adj}} = S$$

and given that fx represents a force per unit mass, the expression becomes:

$$f_{x,g} = gS.$$

6. Assuming the energy grade line is not the same as the channel slope, and for a reach of consistent slope there is a consistent friction loss, it follows that:

$$f_{x,f} = S_f g.$$

7. All of these assumptions combined arrives at the 1-dimensional Saint-Venant equation in the x-direction:

$$\underset{(a)}{\frac{\partial u}{\partial t}} + \underset{(b)}{u\frac{\partial u}{\partial x}} + \underset{(c)}{g\frac{\partial h}{\partial x}} + \underset{(d)}{g(S_f} - \underset{(e)}{S)} = 0,$$

where (a) is the local acceleration term, (b) is the convective acceleration term, (c) is the pressure gradient term, (d) is the friction term, and (e) is the gravity term.

Terms

The local acceleration (a) can also be thought of as the "unsteady term" as this describes some change in velocity over time. The convective acceleration (b) is an acceleration caused by some change in velocity over position, for example the speeding up or slowing down of a fluid entering a constriction or an opening, respectively. Both these terms make up the inertia terms of the 1-dimensional Saint-Venant equation.

The pressure gradient term (c) describes how pressure changes with position, and since the pressure is assumed hydrostatic, this is the change in head over position. The friction term (d) accounts for losses in energy due to friction, while the gravity term (e) is the acceleration due to bed slope.

Wave Modelling by Shallow Water Equations

Shallow water equations can be used to model Rossby and Kelvin waves in the atmosphere, rivers, lakes and oceans as well as gravity waves in a smaller domain (e.g. surface waves in a bath). In order for shallow water equations to be valid, the wavelength of the phenomenon they are supposed to model has to be much higher than the depth of the basin where the phenomenon takes place. Shallow water equations are especially suitable to model tides which have very large length scales (over hundred of kilometers). For tidal motion, even a very deep ocean may be considered as shallow as its depth will always be much smaller than the tidal wavelength.

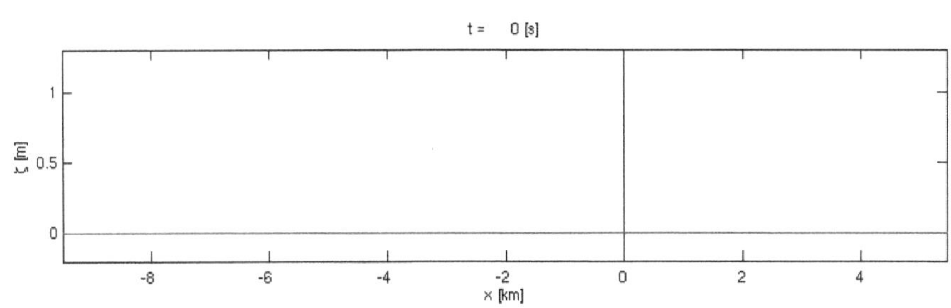

Tsunami generation and propagation, as computed with the shallow water equations (red line; without frequency dispersion)), and with a Boussinesq-type model (blue line; with frequency dispersion). Observe that the Boussinesq-type model (blue line) forms a soliton with an oscillatory tail staying behind. The shallow water equations (red line) form a steep front, which will lead to bore formation, later on. The water depth is 100 meters.

REFERENCES

1. Robert E. Horton, The Horton Papers (1933)
2. Keith Beven, Robert E. Horton's perceptual model of infiltration processes, Hydrological Processes, Wiley Intersciences DOI 10:1002 hyp 5740 (2004)
3. L. Davis Mackenzie and Susan J. Masten, Principles of Environmental Engineering and Science ISBN 0-07-235053-9
4. McMahon T.A. and Finlayson, B.; Global Runoff: Continental Comparisons of Annual Flows and Peak Discharges ISBN 3-923381-27-1
5. Nelson, R. (2004). The Water Cycle. Minneapolis: Lerner. ISBN 0-8225-4596-9
6. United States. National Research Council. Washington, DC. "Urban Stormwater Management in the United States." October 15, 2008. pp. 18-20.
7. Wigley T.M.L & Jones P.D (1985). "Influences of precipitation changes and direct CO2 effects on streamflow". Letters to Nature.
8. W.F. Spencer, Distribution of Pesticides between Soil, Water and Air, International symposium on Pesticides in the Soil, February 25–27, 1970, Michigan State University, East Lansing, Michigan
9. Science News. "DDT treatment turns male fish into mothers." 2000-02-05. (By subscription only.)
10. Klimaszyk Piotr, Rzymski Piotr "Surface Runoff as a Factor Determining Trophic State of Midforest Lake" Polish Journal of Environmental Studies, 2011, 20(5), 1203-1210
11. Center for Neighborhood Technology, Chicago IL "The Prevalence and Cost of Urban Flooding." May 2013 http://www.cnt.org/media/CNT_PrevalenceAndCostOfUrbanFlooding.pdf
12. U.S. Environmental Protection Agency (EPA). "Impervious Cover." Ecosystems Research Division, Athens, GA. 2009-02-24. Archived May 9, 2009, at the Wayback Machine.
13. Urban Runoff, City of Santa Monica website. Retrieved 29 July 2007.
14. Maryland Department of Environment. Baltimore, MD. "Erosion and Sediment Control and Stormwater Management in Maryland." 2007. Archived September 12, 2008, at the Wayback Machine.
15. Channel Stability Assessment for Flood Control Projects U.S. Army Corps of Engineers, (1996) ISBN 0-7844-0201-9
16. United States. Code of Federal Regulations, 40 CFR 122.26

17. EPA. Washington, D.C. "Stormwater Discharges From Municipal Separate Storm Sewer Systems (MS4s)." 2009-03-11.

18. C.M. Hogan, Leda Patmore, Gary Latshaw, Harry Seidman et al. Computer modeling of pesticide transport in soil for five instrumented watersheds, U.S. Environmental Protection Agency Southeast Water laboratory, Athens, Ga. by ESL Inc., Sunnyvale, California (1973)

19. "The Shallow Water Equations" (PDF). Retrieved 2010-01-22.

20. Clint Dawson and Christopher M. Mirabito (2008). "The Shallow Water Equations" (PDF). Retrieved 2013-03-28.

21. S. Néelz and G Pender (2009). "Desktop review of 2D hydraulic modelling packages". Joint Environment Agency/Defra Flood and Coastal Erosion Risk Management Research and Development Programme (Science Report: SC080035): p.5. Retrieved 2 December 2016.

22. Saint-Venant, A. J. C. Barré de (1871), Théorie du mouvement non permanent des eaux, avec application aux crues des rivières et a l'introduction de marées dans leurs lits. Comptes Rendus des Séances de l'Académie des Sciences, vol. 73, pp. 147–154 & 237–240.

23. Chow, Ven Te (1959), Open-channel hydraulics, McGraw-Hill, OCLC 4010975, §18-1 & §18-2.

24. Cunge, J. A., F. M. Holly Jr. and A. Verwey (1980), Practical aspects of computational river hydraulics, Pitman Publishing, ISBN 0 273 08442 9, §§2.1 & 2.2

25. Whitham, G. B. (1974) Linear and Nonlinear Waves, §§5.2 & 13.10, Wiley, ISBN 0-471-94090-9

26. Lighthill, J. (2005), Waves in fluids, Cambridge University Press, ISBN 978-0-521-01045-0, §§2.8–2.14

27. Meyer, R. E. (1960), Theory of characteristics of inviscid gas dynamics. In: Fluid Dynamics/Strömungsmechanik, Encyclopedia of Physics IX, Eds. S. Flügge & C. Truesdell ,Springer, Berlin, ISBN 978-3-642-45946-7, pp. 225–282

28. Didenkulova, I. and E. Pelinovsky (2011), Rogue waves in nonlinear hyperbolic systems (shallow-water framework), Nonlinearity 24(3), pp. R1–R18, doi:10.1088/0951-7715/24/3/R01

29. Brunner, G. W. (1995), HEC-RAS River Analysis System. Hydraulic Reference Manual. Version 1.0 Rep., DTIC Document.

30. Searby, D.; Dean, A.; Margetts J. (1998), Christchurch harbour Hydroworks modelling., Proceedings of the WAPUG Autumn meeting, Blackpool, UK.

31. Havnø, K., M. Madsen, J. Dørge, and V. Singh (1995), MIKE 11-a generalized river modelling package, Computer models of watershed hydrology., 733–782.

32. Yeh, G.; Cheng, J.; Lin, J.; Martin, W. (1995), A numerical model simulating water flow and contaminant and sediment transport in watershed systems of 1-D stream-river network, 2-D overland regime, and 3-D subsurface media . Computer models of watershed hydrology, 733–782.

33. Novak, P., et al., Hydraulic Modelling – An Introduction: Principles, Methods and Applications. 2010: CRC Press.

34. Scharffenberg, W. A., and M. J. Fleming (2006), Hydrologic Modeling System HEC-HMS: User's Manual, US Army Corps of Engineers, Hydrologic Engineering Center.

35. DHI (Danish Hydraulic Institute) (2011), MIKE SHE User Manual Volume 2: Reference Guide, edited.

36. Bates, P., T. Fewtrell, M. Trigg, and J. Neal (2008), LISFLOOD-FP user manual and technical note, code release 4.3. 6, University of Bristol.

CHAPTER 8

Groundwater

Chapter Contents

- **Introduction**
- **Aquifers**
- **Water Cycle**
- **Issues**

Groundwater is water that is located below the earth's surface. Over time, water from rain and rivers migrates through the ground and is stored in porous soils and rocks. The study of groundwater is known as hydrogeology.

Groundwater is found in vast quantities filling the spaces between grains of soil or rock; it slowly flows through aquifers; it connects with rivers, streams, lakes and wetlands; it feeds trees and vegetation. Australia is a very old continent, and much of its groundwater is tens of thousands, even hundreds of thousands of years old.

INTRODUCTION

Groundwater (or ground water) is the water present beneath Earth's surface in soil pore spaces and in the fractures of rock formations. A unit of rock or an unconsolidated deposit is called an aquifer when it can yield a usable quantity of water. The depth at which soil pore spaces or fractures and voids in rock become completely saturated with water is called the water table. Groundwater is recharged from, and eventually flows to, the surface naturally; natural discharge often occurs at springs and seeps, and can form oases or wetlands.

Groundwater is also often withdrawn for agricultural, municipal, and industrial use by constructing and operating extraction wells. The study of the distribution and movement of groundwater is hydrogeology, also called groundwater hydrology.

Typically, groundwater is thought of as water flowing through shallow aquifers, but, in the technical sense, it can also contain soil moisture, permafrost (frozen soil), immobile water in very low permeability bedrock, and deep geothermal or oil formation water. Groundwater is hypothesized to provide lubrication that can possibly influence the movement of faults. It is likely that much of Earth's subsurface contains some water, which may be mixed with other fluids in some instances. Groundwater may not be confined only to Earth. The formation of some of the landforms observed on Mars may have been influenced by groundwater. There is also evidence that liquid water may also exist in the subsurface of Jupiter's moon Europa.

The entire surface water flow of the Alapaha River near Jennings, Florida going into a sinkhole leading to the Floridan Aquifer groundwater

Groundwater is often cheaper, more convenient and less vulnerable to pollution than surface water. Therefore, it is commonly used for public water supplies. For example, groundwater provides the largest source of usable water storage in the United States, and California annually withdraws the largest amount of groundwater of all the states. Underground reservoirs contain far more water than the capacity of all surface reservoirs and lakes in the US, including the Great Lakes. Many municipal water supplies are derived solely from groundwater.

Polluted groundwater is less visible, but more difficult to clean up, than pollution in rivers and lakes. Groundwater pollution most often results from improper disposal of wastes on land. Major sources include industrial and household chemicals and garbage landfills, excessive fertilizers and pesticides used in agriculture, industrial waste lagoons, tailings and process wastewater from mines, industrial fracking, oil field brine pits, leaking underground oil storage tanks and pipelines, sewage sludge and septic systems.

AQUIFERS

An *aquifer* is a layer of porous substrate that contains and transmits groundwater. When water can flow directly between the surface and the saturated zone of an aquifer, the aquifer is unconfined. The deeper parts of unconfined aquifers are usually more saturated since gravity causes water to flow downward.

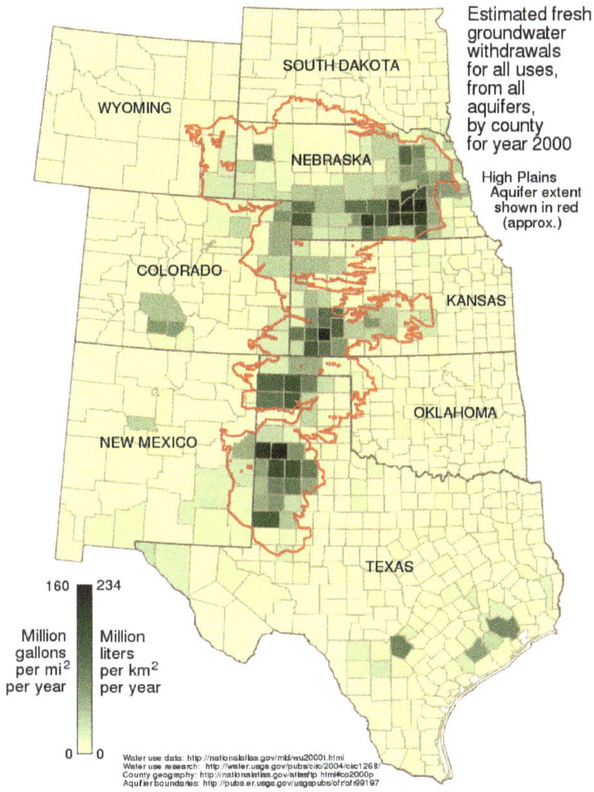

Groundwater withdrawal rates from the Ogallala Aquifer in the Central United States

The upper level of this saturated layer of an unconfined aquifer is called the *water table* or *phreatic surface*. Below the water table, where in general all pore spaces are saturated with water, is the phreatic zone.

Substrate with low porosity that permits limited transmission of groundwater is known as an *aquitard*. An *aquiclude* is a substrate with porosity that is so low it is virtually impermeable to groundwater.

A *confined aquifer* is an aquifer that is overlain by a relatively impermeable layer of rock or substrate such as an aquiclude or aquitard. If a confined aquifer follows a downward grade from its *recharge zone*, groundwater can become pressurized as it flows. This can create artesian wells that flow freely without the need of a pump and rise to a higher elevation than the static water table at the above, unconfined, aquifer.

The characteristics of aquifers vary with the geology and structure of the substrate and topography in which they occur. In general, the more productive aquifers occur in sedimentary geologic formations. By comparison, weathered and fractured crystalline rocks yield smaller quantities of groundwater in many environments. Unconsolidated to poorly cemented alluvial materials that have accumulated as valley-filling sediments in major river valleys and geologically subsiding structural basins are included among the most productive sources of groundwater.

The high specific heat capacity of water and the insulating effect of soil and rock can mitigate the effects of climate and maintain groundwater at a relatively steady temperature. In some places where groundwater temperatures are maintained by this effect at about 10 °C (50 °F), groundwater can be used for controlling the temperature inside structures at the surface. For example, during hot weather relatively cool groundwater can be pumped through radiators in a home and then returned to the ground in another well. During cold seasons, because it is relatively warm, the water can be used in the same way as a source of heat for heat pumps that is much more efficient than using air.

The volume of groundwater in an aquifer can be estimated by measuring water levels in local wells and by examining geologic records from well-drilling to determine the extent, depth and thickness of water-bearing sediments and rocks. Before an investment is made in production wells, test wells may be drilled to measure the depths at which water is encountered and collect samples of soils, rock and water for laboratory analyses. Pumping tests can be performed in test wells to determine flow characteristics of the aquifer.

WATER CYCLE

Groundwater makes up about twenty percent of the world's fresh water supply, which is about 0.61% of the entire world's water, including oceans and permanent ice. Global groundwater storage is roughly equal to the total amount of freshwater stored in the snow and ice pack, including the north and south poles. This makes it an important resource that can act as a natural storage that can buffer against shortages of surface water, as in during times of drought.

Groundwater is naturally replenished by surface water from precipitation, streams, and rivers when this recharge reaches the water table.

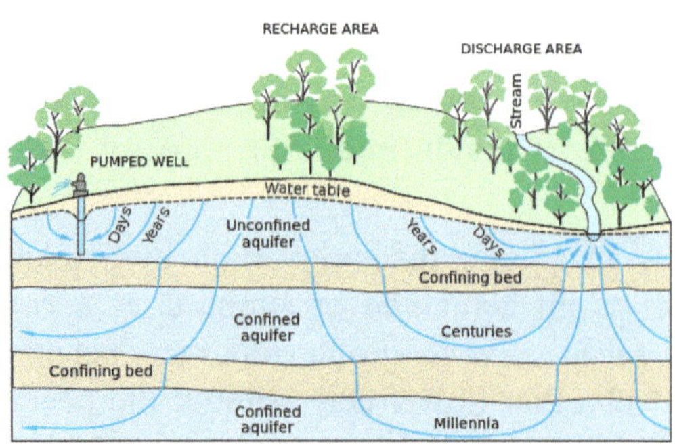

Relative groundwater travel times

Dzherelo, a common source of drinking water in a Ukrainian village

Groundwater can be a long-term 'reservoir' of the natural water cycle (with residence times from days to millennia), as opposed to short-term water reservoirs like the atmosphere and fresh surface water (which have residence times from minutes to years). The figure shows how deep groundwater (which is quite distant from the surface recharge) can take a very long time to complete its natural cycle.

The Great Artesian Basin in central and eastern Australia is one of the largest confined aquifer systems in the world, extending for almost 2 million km^2. By analyzing the trace elements in water sourced from deep underground, hydrogeologists have been able to determine that water extracted from these aquifers can be more than 1 million years old.

By comparing the age of groundwater obtained from different parts of the Great Artesian Basin, hydrogeologists have found it increases in age across the basin. Where water recharges the aquifers along the Eastern Divide, ages are young. As groundwater flows westward across the continent, it increases in age, with the oldest groundwater occurring in the western parts. This means that in order to have travelled almost 1000 km from the source of recharge in 1 million years, the groundwater flowing through the Great Artesian Basin travels at an average rate of about 1 meter per year.

Reflective carpet trapping soil water vapor

Recent research has demonstrated that evaporation of groundwater can play a significant role in the local water cycle, especially in arid regions. Scientists in Saudi Arabia have proposed plans to recapture and recycle this evaporative moisture for crop irrigation. In the opposite photo, a 50-centimeter-square reflective carpet, made of small adjacent plastic cones, was placed in a plant-free dry desert area for five months, without rain or irrigation. It managed to capture and condense enough ground vapor to bring to life naturally buried seeds underneath it, with a green area of about 10% of the carpet area. It is expected that, if seeds were put down before placing this carpet, a much wider area would become green.

ISSUES

Overview

Certain problems have beset the use of groundwater around the world. Just as river waters have been over-used and polluted in many parts of the world, so too have aquifers. The big difference is that aquifers are out of sight. The other major problem is that water management agencies, when calculating the "sustainable yield" of aquifer and river water, have often counted the same water twice, once in the aquifer, and once in its connected river. This problem, although understood for centuries, has persisted, partly through inertia within government agencies. In Australia, for example, prior to the statutory reforms initiated by the Council of Australian Governments water reform framework in the 1990s, many Australian states managed groundwater and surface water through separate government agencies, an approach beset by rivalry and poor communication.

In general, the time lags inherent in the dynamic response of groundwater to development have been ignored by water management agencies, decades after scientific understanding of the issue was consolidated. In brief, the effects of groundwater overdraft (although undeniably real) may take decades or centuries to manifest themselves. In a classic study in 1982, Bredehoeft and colleagues modeled a situation where groundwater extraction in an intermontane basin withdrew the entire annual recharge, leaving 'nothing' for the natural groundwater-dependent vegetation community. Even when the borefield was situated close to the vegetation, 30% of the original vegetation demand could still be met by the lag inherent in the system after 100 years. By year 500, this had reduced to 0%, signalling complete death of the groundwater-dependent vegetation. The science has been available to make these calculations for decades; however, in general water management agencies have ignored effects that will appear outside the rough timeframe of political elections (3 to 5 years). Marios Sophocleous argued strongly that management agencies must define and use appropriate timeframes in groundwater planning. This will mean calculating groundwater withdrawal permits based on predicted effects decades, sometimes centuries in the future.

As water moves through the landscape, it collects soluble salts, mainly sodium chloride. Where such water enters the atmosphere through evapotranspiration, these salts are left behind. In irrigation districts, poor drainage of soils and surface aquifers can result in water tables' coming to the surface in low-lying areas. Major land degradation problems of soil salinity and waterlogging result, combined with increasing levels of salt in surface waters. As a consequence, major damage has occurred to local economies and environments.

Four important effects are worthy of brief mention. First, flood mitigation schemes, intended to protect infrastructure built on floodplains, have had the unintended consequence of reducing aquifer recharge associated with natural flooding. Second, prolonged depletion of groundwater in extensive aquifers can result in land subsidence, with associated infrastructure damage – as well as, third, saline intrusion. Fourth, draining acid sulphate soils, often found in low-lying coastal plains, can result in acidification and pollution of formerly freshwater and estuarine streams.

Another cause for concern is that groundwater drawdown from over-allocated aquifers has the potential to cause severe damage to both terrestrial and aquatic ecosystems – in some cases very conspicuously but in others quite imperceptibly because of the extended period over which the damage occurs.

Overdraft

Groundwater is a highly useful and often abundant resource. However, over-use, or overdraft, can cause major problems to human users and to the environment. The most evident problem (as far as human groundwater use is concerned) is a lowering of the water table beyond the reach of existing wells. As a consequence, wells must be drilled deeper to reach the groundwater; in some places (e.g., California, Texas, and India) the water table has dropped hundreds of feet because of extensive well pumping. In the Punjab region of India, for example, groundwater levels have dropped 10 meters since 1979, and the rate of depletion is accelerating. A lowered water table may, in turn, cause other problems such as groundwater-related subsidence and saltwater intrusion.

Wetlands contrast the arid landscape around Middle Spring, Fish Springs National Wildlife Refuge, Utah

Groundwater is also ecologically important. The importance of groundwater to ecosystems is often overlooked, even by freshwater biologists and ecologists. Groundwaters sustain rivers, wetlands, and lakes, as well as subterranean ecosystems within karst or alluvial aquifers.

Not all ecosystems need groundwater, of course. Some terrestrial ecosystems – for example, those of the open deserts and similar arid environments – exist on irregular rainfall and the moisture it delivers to the soil, supplemented by moisture in the air. While there are other terrestrial ecosystems in more hospitable environments where groundwater plays no central role, groundwater is in fact fundamental to many of the world's major ecosystems. Water flows between groundwaters and surface waters. Most rivers, lakes, and wetlands are fed by, and (at other places or times) feed groundwater, to varying degrees. Groundwater feeds soil moisture through percolation, and many terrestrial vegetation communities depend directly on either groundwater or the percolated soil moisture above the aquifer for at least part of each year. Hyporheic zones (the mixing zone of streamwater and groundwater) and riparian zones are examples of ecotones largely or totally dependent on groundwater.

Subsidence

Subsidence occurs when too much water is pumped out from underground, deflating the space below the above-surface, and thus causing the ground to collapse. The result can look like craters on plots of land. This occurs because, in its natural equilibrium state, the hydraulic pressure of groundwater in the pore spaces of the aquifer and the aquitard supports some of the weight of the overlying sediments. When groundwater is removed from aquifers by excessive pumping, pore pressures in the aquifer drop and compression of the aquifer may occur. This compression may be partially recoverable if pressures rebound, but much of it is not. When the aquifer gets compressed, it may cause land subsidence, a drop in the ground surface. The city of New Orleans, Louisiana is actually below sea level today, and its subsidence is partly caused by removal of groundwater from the various aquifer/aquitard systems beneath it. In the first half of the 20th century, the San Joaquin Valley experienced significant subsidence, in some places up to 8.5 metres (28 feet) due to groundwater removal. Cities on river deltas, including Venice in Italy,and Bangkok in Thailand, have experienced surface subsidence; Mexico City, built on a former lake bed, has experienced rates of subsidence of up to 40 cm (1'3") per year.

Seawater Intrusion

In general, in very humid or undeveloped regions, the shape of the water table mimics the slope of the surface. The recharge zone of an aquifer near the seacoast is likely to be inland, often at considerable distance. In these coastal areas, a lowered water table may induce sea water to reverse the flow toward the land. Sea water moving inland is called a saltwater intrusion. In alternative fashion, salt from mineral beds may leach into the groundwater of its own accord.

Pollution

Polluted groundwater is less visible, but more difficult to clean up, than pollution in rivers and lakes. Groundwater pollution most often results from improper disposal of wastes on land. Major sources include industrial and household chemicals and garbage landfills, industrial waste lagoons, tailings and process wastewater from mines, oil field brine pits, leaking underground oil storage tanks and pipelines, sewage sludge and septic systems. Polluted groundwater is mapped by sampling soils and groundwater near suspected or known sources of pollution, to determine the extent of the pollution, and to aid in the design of groundwater remediation systems. Preventing groundwater pollution near potential sources such as landfills requires lining the bottom of a landfill with watertight materials, collecting any leachate with drains, and keeping rainwater off any potential contaminants, along with regular monitoring of nearby groundwater to verify that contaminants have not leaked into the groundwater.

Iron oxide staining caused by reticulation from an unconfined aquifer in karst topography. Perth, Western Australia.

Groundwater pollution, from pollutants released to the ground that can work their way down into groundwater, can create a contaminant plume within an aquifer. Pollution can occur from landfills, naturally occurring arsenic, on-site sanitation systems or other point sources, such as petrol stations or leaking sewers.

Movement of water and dispersion within the aquifer spreads the pollutant over a wider area, its advancing boundary often called a plume edge, which can then intersect with groundwater wells or daylight into surface water such as seeps and springs, making the water supplies unsafe for humans and wildlife. Different mechanism have influence on the transport of pollutants, e.g. diffusion, adsorption, precipitation, decay, in the groundwater. The interaction of groundwater contamination with surface waters is analyzed by use of hydrology transport models.

The danger of pollution of municipal supplies is minimized by locating wells in areas of deep groundwater and impermeable soils, and careful testing and monitoring of the aquifer and nearby potential pollution sources.

Arsenic and Fluoride

Around one-third of the world's population drinks water from groundwater resources. Of this, about 10 percent, approximately 300 million people, obtains water from groundwater resources that are heavily polluted with arsenic or fluoride. These trace elements derive mainly from natural sources by leaching from rock and sediments.

New Method of Identifying Substances That are Hazardous To Health

In 2008, the Swiss Aquatic Research Institute, Eawag, presented a new method by which hazard maps could be produced for geogenic toxic substances in groundwater. This provides an efficient way of determining which wells should be tested.

In 2016, the research group made its knowledge freely available on the Groundwater Assessment Platform GAP. This offers specialists worldwide the possibility of uploading their own measurement data, visually displaying them and producing risk maps for areas of their choice. GAP also serves as a knowledge-sharing forum for enabling further development of methods for removing toxic substances from water.

Government Regulations

In the United States, laws regarding ownership and use of groundwater are generally state laws; however, regulation of groundwater to minimize pollution of groundwater is by both states and the federal-level Environmental Protection Agency. Ownership and use rights to groundwater typically follow one of three main systems:

Rule of Capture

The Rule of Capture provides each landowner the ability to capture as much groundwater as they can put to a beneficial use, but they are not guaranteed any set amount of water. As a result, well-owners are not liable to other landowners for taking water from beneath their land. State laws or regulations will often define "beneficial use", and sometimes place other limits, such as disallowing groundwater extraction which causes subsidence on neighboring property.

Riparian Rights

Limited private ownership rights similar to riparian rights in a surface stream. The amount of groundwater right is based on the size of the surface area where each landowner gets a corresponding amount of the available water. Once adjudicated, the maximum amount of the water right is set, but the right can be decreased if the total amount of available water decreases as is likely during a drought. Landowners may sue others for encroaching upon their groundwater rights, and water pumped for use on the overlying land takes preference over water pumped for use off the land.

Environmental Protection of Groundwater

In November 2006, the Environmental Protection Agency published the groundwater Rule in the United States Federal Register. The EPA was worried that the groundwater system would be vulnerable to contamination from fecal matter. The point of the rule was to keep microbial pathogens out of public water sources. The 2006 groundwater Rule was an amendment of the 1996 Safe Drinking Water Act.

Others

Reasonable Use Rule (American Rule)

This rule does not guarantee the landowner a set amount of water, but allows unlimited extraction as long as the result does not unreasonably damage other wells or the aquifer system. Usually this rule gives great weight to historical uses and prevents new uses that interfere with the prior use.

Groundwater Scrutiny Upon Real Estate Property Transactions in the US

In the US, upon commercial real estate property transactions both groundwater and soil are the subjects of scrutiny, with a Phase I Environmental Site Assessment normally being prepared to investigate and disclose potential pollution issues. In the San Fernando Valley of California, real estate contracts for property transfer below the Santa Susana Field Laboratory (SSFL) and eastward have clauses releasing the seller from liability for groundwater contamination consequences from existing or future pollution of the Valley Aquifer.

REFERENCES

1. Richard Greenburg (2005). The Ocean Moon: Search for an Alien Biosphere. Springer Praxis Books.
2. National Geographic Almanac of Geography, 2005, ISBN 0-7922-3877-X, page 148.
3. "What is hydrology and what do hydrologists do?". The USGS Water Science School. United States Geological Survey. 23 May 2013. Retrieved 21 Jan 2014.
4. "Learn More: Groundwater". Columbia Water Center. Retrieved 15 September 2009.
5. United States Department of the Interior (1977). Ground Water Manual (First ed.). United States Government Printing Office. p. 4.

6. File:Groundwater flow.svg
7. Hassan, SM Tanvir (March 2008). Assessment of groundwater evaporation through groundwater model with spatio-temporally variable fluxes (PDF) (MSc). Enschede, Netherlands: International Institute for Geo-Information Science and Earth Observation.
8. Al-Kasimi, S. M. (2002). Existence of Ground Vapor-Flux Up-Flow: Proof & Utilization in Planting The Desert Using Reflective Carpet. 3. Dahran. pp. 105–119.
9. Sophocleous, Marios (2002). "Interactions between groundwater and surface water: the state of the science". Hydrogeology Journal. 10: 52–67. Bibcode:2002HydJ...10...52S. doi:10.1007/s10040-001-0170-8.
10. "Free articles and software on drainage of waterlogged land and soil salinity control". Retrieved 2010-07-28.
11. Ludwig, D.; Hilborn, R.; Walters, C. (1993). "Uncertainty, Resource Exploitation, and Conservation: Lessons from History" (PDF). Science. 260 (5104): 17–36. Bibcode:1993Sci...260...17L. doi:10.1126/science.260.5104.17. JSTOR 1942074. PMID 17793516.
12. Zektser et al.
13. Sommer, Bea; Horwitz, Pierre; Sommer, Bea; Horwitz, Pierre (2001). "Water quality and macroinvertebrate response to acidification following intensified summer droughts in a Western Australian wetland". Marine and Freshwater Research. 52 (7): 1015. doi:10.1071/MF00021.
14. Zektser, S.; Lolciga, H. A.; Wolf, J. T. (2004). "Environmental impacts of groundwater overdraft: selected case studies in the southwestern United States". Environmental Geology. 47 (3): 396–404. doi:10.1007/s00254-004-1164-3. replacement character in |last2= at position 3 (help)
15. Upmanu Lall. "Punjab: A tale of prosperity and decline". Columbia Water Center. Retrieved 2009-09-11.
16. Dokka, Roy K. (2011). "The role of deep processes in late 20th century subsidence of New Orleans and coastal areas of southern Louisiana and Mississippi". Journal of Geophysical Research. 116 (B6). doi:10.1029/2010JB008008. ISSN 0148-0227.
17. Sneed, M; Brandt, J; Solt, M (2013). "Land Subsidence along the Delta-Mendota Canal in the Northern Part of the San Joaquin Valley, California, 2003–10" (PDF). USGS Scientific Investigations Report 2013-5142. Retrieved 22 June 2015.
18. Tosi, Luigi; Teatini, Pietro; Strozzi, Tazio; Da Lio, Cristina (2014). "Relative Land Subsidence of the Venice Coastland, Italy": 171–173. doi:10.1007/978-3-319-08660-6_32.
19. Aobpaet, Anuphao; Cuenca, Miguel Caro; Hooper, Andrew; Trisirisatayawong, Itthi (2013). "InSAR time-series analysis of land subsidence in Bangkok, Thailand". International Journal of Remote Sensing. 34 (8): 2969–2982. doi:10.1080/01431161.2012.756596. ISSN 0143-1161.

20. Arroyo, Danny; Ordaz, Mario; Ovando-Shelley, Efrain; Guasch, Juan C.; Lermo, Javier; Perez, Citlali; Alcantara, Leonardo; Ramírez-Centeno, Mario S. (2013). "Evaluation of the change in dominant periods in the lake-bed zone of Mexico City produced by ground subsidence through the use of site amplification factors". Soil Dynamics and Earthquake Engineering. 44: 54–66. doi:10.1016/j.soildyn.2012.08.009. ISSN 0267-7261.
21. Eawag (2015) Geogenic Contamination Handbook – Addressing Arsenic and Fluoride in Drinking Water. C.A. Johnson, A. Bretzler (Eds.), Swiss Federal Institute of Aquatic Science and Technology (Eawag), Duebendorf, Switzerland. (download: www.eawag.ch/en/research/humanwelfare/drinkingwater/wrq/geogenic-contamination-handbook/)
22. Amini, M.; Mueller, K.; Abbaspour, K.C.; Rosenberg, T.; Afyuni, M.; Møller, M.; Sarr, M.; Johnson, C.A. (2008) Statistical modeling of global geogenic fluoride contamination in groundwaters. Environmental Science and Technology, 42(10), 3662-3668, doi:10.1021/es071958y
23. Amini, M.; Abbaspour, K.C.; Berg, M.; Winkel, L.; Hug, S.J.; Hoehn, E.; Yang, H.; Johnson, C.A. (2008). "Statistical modeling of global geogenic arsenic contamination in groundwater". Environmental Science and Technology 42 (10), 3669-3675. doi:10.1021/es702859e
24. Winkel, L.; Berg, M.; Amini, M.; Hug, S.J.; Johnson, C.A. Predicting groundwater arsenic contamination in Southeast Asia from surface parameters. Nature Geoscience, 1, 536–542 (2008). doi:10.1038/ngeo254
25. Rodríguez-Lado, L.; Sun, G.; Berg, M.; Zhang, Q.; Xue, H.; Zheng, Q.; Johnson, C.A. (2013) Groundwater arsenic contamination throughout China. Science, 341(6148), 866-868, doi:10.1126/science.1237484
26. "Appendix H, Groundwater Law and Regulated Riparianism", Final Report: Restoring Great Lakes Basin Water thorough the Use of Conservation Credits and Integrated Water Balance Analysis System, The Great Lakes Protection Fund Project # 763 (pdf), retrieved 16 January 2014
27. Ground Water Rule (GWR) | Ground Water Rule | US EPA. Water.epa.gov. Retrieved on 2011-06-09.
28. EPA; http://water.epa.gov/type/groundwater/index

CHAPTER 9

Watershed Management

Chapter Contents

Watershed management is an adaptive, comprehensive, integrated multi-resource management planning process that seeks to balance healthy ecological, economic, and cultural/social conditions within a watershed. Watershed management serves to integrate planning for land and water; it takes into account both ground and surface water flow, recognizing and planning for the interaction of water, plants, animals and human land use found within the physical boundaries of a watershed.

Watershed management provides a framework for integrated decision-making to help: assess the nature and status of the watershed; identify watershed issues; define and re-evaluate short and long-term objectives, actions and goals; assess benefits and costs; and implement and evaluate actions.

INTRODUCTION

Watershed management is the study of the relevant characteristics of a watershed aimed at the sustainable distribution of its resources and the process of creating and implementing plans, programs, and projects to sustain and enhance watershed functions that affect the plant, animal, and human communities within a watershed boundary. Features of a watershed that agencies seek to manage include water supply, water quality, drainage, stormwater runoff, water rights, and the overall planning and utilization of watersheds. Landowners, land use agencies, stormwater management experts, environmental specialists, water use surveyors and communities all play an integral part in watershed management.

OBJECTIVES OF WATERSHED MANAGEMENT

The different objectives of watershed management programs are:

- To protect, conserve and improve the land of watershed for more efficient and sustained production.
- To protect and enhance the water resource originating in the watershed.
- To check soil erosion and to reduce the effect of sediment yield on the watershed.
- To rehabilitate the deteriorating lands.
- To moderate the floods peaks at down stream areas.
- To increase infiltration of rainwater.
- To improve and increase the production of timbers, fodder and wild life resource.
- To enhance the ground water recharge, wherever applicable.
- To reduce the occurrence of floods and the resultant damage by adopting strategies for flood management.
- To provide standard quality of water by encouraging vegetation and waste disposal facilities.

NEED OF WATERSHED MANAGEMENT

An integrated watershed management approach needs to be adopted and the soil and water conservation technologies and approaches need to be applied in field situations by the officer-trainees. The Indian Institute of Soil and Water Conservation (IISWC) director Pravin Shinagare said this while addressing a multi-disciplinary team of 20 officers from Odisha at the conclusion of a five-day training programme on soil and water conservation training-cum-exposure visit at the institute on Friday. The trainees were sponsored by the Institute on Management of Agricultural Extension, Bhubaneswar.

Mishra spoke on the concept, philosophy, importance of conserving natural resources through integrated watershed management. He advised the participants to give more thought on this mechanism. Speaking on the occasion, Plant Science head OP Chaturvedi stressed on importance of work in close collaboration with people and different agencies and not in isolation. He said that common property resources should be protected, conserved and utilised with community participation for their common cause and development. The HRD and Social Science division head Lakhan Singh motivated the officers to use past experiences and relate with existing agro-ecosystems. He highlighted the various factors promoting and inhibiting people's participation in watershed management. He said the commitment of villagers and officers towards watershed goal will make a difference in socio-economic transformation of people.

Watersheds sustain life, in more ways than one. According to the Environmental Protection Agency, more than **$450 billion** in foods, fiber, manufactured goods and tourism depend on clean, healthy watersheds. That is why proper watershed protection is necessary to you and your community. Watershed protection is a means of protecting a lake, river, or stream by managing the entire watershed that drains into it. Clean, healthy watersheds depend on an informed public to make the right decisions when it comes to the environment and actions made by the community.

Watershed management practices in terms of purpose

- To increase infiltration
- To increase water holding capacity
- To prevent soil erosion
- Method and accomplishment

In brief various control measures are:

- Vegetative measures (Agronomical measures)
- Strip cropping
- Pasture cropping
- Grass land farming
- Wood lands
- Engineering measures (Structural practices)
- Contour bunding
- contour trenching
- Terracing
- Construction of earthen embankment
- Construction of check dams
- Construction of farm ponds
- Construction of diversion
- Gully controlling structure
- Rock dam
- Establishment of permanent grass and vegetation
- Providing vegetative and stone barriers
- Construction of silt tanks dentension

Influence of soil conservation measures and vegetation cover on erosion, Runoff and Nutrient loss.

Controlling Pollution

In agricultural systems, common practices include the use of buffer strips, grassed waterways, the reestablishment of wetlands, and forms of sustainable agriculture practices such as conservation tillage, crop rotation and intercropping. After certain practices are installed, it is important to continually monitor these systems to ensure that they are working properly in terms of improving environmental quality.

In urban settings, managing areas to prevent soil loss and control stormwater flow are a few of the areas that receive attention. A few practices that are used to manage stormwater before it reaches a channel are retention ponds, filtering systems and wetlands. It is important that stormwater is given an opportunity to infiltrate so that the soil and vegetation can act as a "filter" before the water reaches nearby streams or lakes. In the case of soil erosion prevention, a few common practices include the use of silt fences, landscape fabric with grass seed and hydroseeding. The main objective in all cases is to slow water movement to prevent soil transport.

Governance

The 2nd World Water Forum held in The Hague in March 2000 raised some controversies that exposed the multilateral nature and imbalance the demand and supply management of freshwater. While donor organisations, private and government institutions backed by the World Bank, believe that freshwater should be governed as an economic good by appropriate pricing, NGOs however, held that freshwater resources should be seen as a social good. The concept of network governance where all stakeholders form partnerships and voluntarily share ideas towards forging a common vision can be used to resolve this clash of opinion in freshwater management. Also, the implementation of any common vision presents a new role for NGOs because of their unique capabilities in local community coordination, thus making them a valuable partner in network governance.

Watersheds replicate this multilateral terrain with private industries and local communities interconnected by a common watershed. Although these groups share a common ecological space that could transcend state borders, their interests, knowledge and use of resources within the watershed are mostly disproportionate and divergent, resulting to the activities of a specific group adversely impacting on other groups. Examples being the Minamata Bay poisoning that occurred from 1932 to 1968, killing over 1,784 individuals and the Wabigoon River incidence of 1962. Furthermore, while some knowledgeable groups are shifting from efficient water resource exploitation to efficient utilization, net gain for the watershed ecology could be lost when other groups seizes the opportunity to exploit more resources. This gap in cooperative communication among multilateral stakeholders within an interconnected watershed, even with the likely presence of the usually reactive and political boundary-constraint state regulations, makes it necessary for the institutionalization of an ecological-scale cooperative network of stakeholders. This concept supports an integrated management style for interconnected natural resources; resonating strongly with the Integrated Water Resources Management system proposed by Global Water Partnership.

Moreover, the need to create partnerships between donor organizations, private and government institutions and community representatives like NGOs in watersheds is to enhance an "organizational society" among stakeholders. This posits a type of public-private partnership, commonly referred to as Type II partnership, which essentially brings together stakeholders that share a common watershed under a voluntary, idea sharing and collectively agreed vision aimed at granting mutual benefits to all stakeholders. Also, it explicates the concept of network governance, which is "the only alternative for collective action", requiring government to rescale its role in decision making and collaborate with other stakeholders on a level playing field rather than in an administrative or hierarchical manner.

Several riparian states have adopted this concept in managing the increasingly scarce resources of watersheds. These include, the nine Rhine states, with a common vision of pollution control, the Lake Chad and river Nile Basins, whose common vision is to ensure environmental sustainability. As a partner in the commonly shared vision, NGOs has adopted a new role in operationalizing the implementation of regional watershed management policies at the local level. For instance, essential local coordination and education are areas where the services of NGOs have been effective. This makes NGOs the "nuclei" for successful watershed management.

ENVIRONMENTAL LAW

Environmental laws often dictate the planning and actions that agencies take to manage watersheds. Some laws require that planning be done, others can be used to make a plan legally enforceable and others set out the ground rules for what can and cannot be done in development and planning. Most countries and states have their own laws regarding watershed management.

Those concerned about aquatic habitat protection have a right to participate in the laws and planning processes that affect aquatic habitats. By having a clear understanding of whom to speak to and how to present the case for keeping our waterways clean a member of the public can become an effective watershed protection advocate.

REFERENCES

1. California Watershed Program
2. Oweyegha-Afunaduula, F.C., I. Afunaduula and M. Balunywa (2003). NGO-sing the Nile Basin Initiative: a myth or reality? Paper at 3rd World Water Forum, Japan, March 2003
3. Evans J: Environmental governance (2011),Routledge, Chapter 4.
4. Mark L, Mark S, John T and Mihriye M (2002); Watershed Partnerships and the Emergence of Collective Action Institutions American Journal of Political Science, Volume 46 No.1 Page 148-163. Available at: http://www.jstor.org/stable/3088419
5. Ewalt, J. G. 2001. Theories of Governance and New Public Management: Links to Understanding Welfare Policy Implementation. Paper Presented at the Annual Conference of the American Society for Public Administration. Also available at: http://unpan1.un.org/intradoc/groups/public/documents/aspa/unpan000563.pdf.
6. A. Emma T and Lisa-Ann H (2010); Public-Private Partnerships for Storm Risk Management in the Cayman Islands; Sustainability Research Institute, The University of Leeds. Also available at: http://www.see.leeds.ac.uk/research/sri/working_papers/SRIPs-21.pdf

7. Milward H and Provan K (2000). "How Networks Are Governed." In Heinrich C and Lynn L eds. Governance and Performance: Models, Methods and Results, Page 243. Washington, DC: Georgetown University Press.

8. Transboundary River Basin Management Regimes: the Rhine basin case study, Newater, pp1-37. Also available at: http://www.tudelft.nl/live/binaries/9229ebc0-66d0-47ca-9d25-5ab2184c85f4/doc/D131_Rhine_Final.pdf

9. "Nile Basin Initiative".

10. Prabhakar K, Lavanya K, and Papa Rao A (2010); NGOs and Farmers Participation in Watershed Development Programme in Prakasam District, Asia-Pacific Journal of Social Science, Volume II (1) Pages173-182. Also available at: http://www.socialsciences-ejournal.org/3.9.%20Prabhakar.pdf.

11. A. Evans J: Environmental governance (2011),Routledge, Chapter 4.

CHAPTER 10

Drainage Basin

Chapter Contents

Drainage basin is an area from which all precipitation flows to a single stream or set of streams. For example, the total area drained by the Mississippi River constitutes its drainage basin, whereas that part of the Mississippi River drained by the Ohio River is the Ohio's drainage basin. The boundary between drainage basins is a drainage divide: all the precipitation on opposite sides of a drainage divide will flow into different drainage basins.

INTRODUCTION

A drainage basin or catchment basin is an extent or an area of land where all surface water from rain, melting snow, or ice converges to a single point at a lower elevation, usually the exit of the basin, where the waters join another body of water, such as a river, lake, reservoir, estuary, wetland, sea, or ocean. Thus if a tributary stream joins a brook that in turn joins a small river which is a tributary of a larger river, there is a series of successively larger (and lower elevation) drainage basins. For instance, the Missouri and Ohio rivers are within their own drainage basins and also within the drainage basin of the Mississippi River.

Other terms used to describe drainage basins are catchment, catchment area, drainage area, river basin and water basin. In North America, the term watershed is commonly used to mean a drainage basin, though in other English-speaking countries, it is used only in its original sense, to mean a drainage divide, the former meaning an area, the latter the high elevation perimeter of that area. Drainage basins drain into other drainage basins in a hierarchical pattern, with smaller sub-drainage basins combining into larger drainage basins.

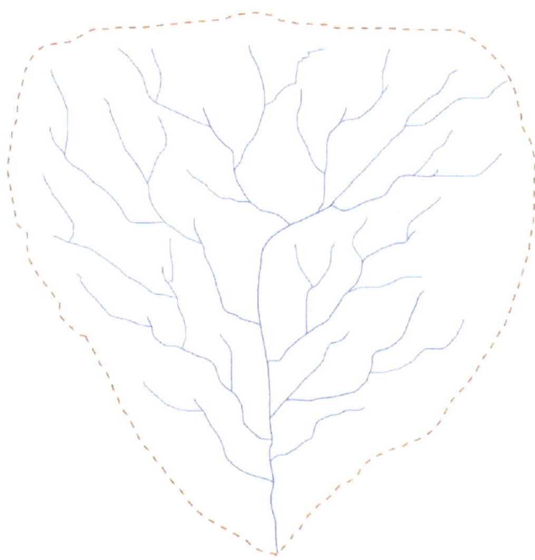

Example of a drainage basin. The dashed line is the main water divide of the hydrographic basin.

Animation of Latorita River drainage basin, Romania

In closed ("endorheic") drainage basins the water converges to a single point inside the basin, known as a sink, which may be a permanent lake, a dry lake, or a point where surface water is lost underground. The drainage basin includes all the streams and rivers that convey the water towards the sink, as well as the land surfaces from which water drains into those channels.

The drainage basin acts as a funnel by collecting all the water within the area covered by the basin and channelling it to a single point. Each drainage basin is separated topographically from adjacent basins by a perimeter, the drainage divide, making up a succession of higher geographical features (such as a ridge, hill or mountains) forming a barrier.

Drainage basins are similar but not identical to hydrologic units, which are drainage areas delineated so as to nest into a multi-level hierarchical drainage system. Hydrologic units are defined to allow multiple inlets, outlets, or sinks. In a strict sense, all drainage basins are hydrologic units but not all hydrologic units are drainage basins.

MAJOR DRAINAGE BASINS OF THE WORLD

Map

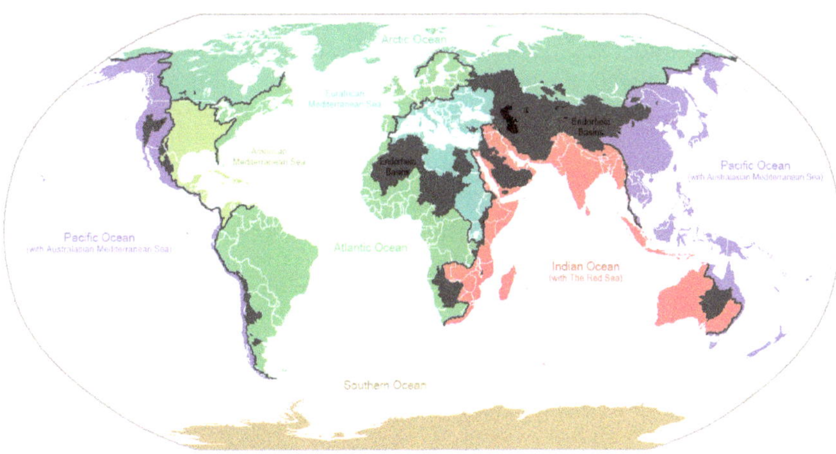

Drainage basins of the principal oceans and seas of the world. Grey areas are endorheic basins that do not drain to the ocean.

Ocean Basins

The following is a list of the major ocean basins:

- About 48.7% of the world's land drains to the Atlantic Ocean. In North America, surface water drains to the Atlantic via the Saint Lawrence River and Great Lakes basins, the Eastern Seaboard of the United States, the Canadian Maritimes, and most of Newfoundland and Labrador. Nearly all of South America east of the Andes also drains to the Atlantic, as does most of Western and Central Europe and the greatest portion of western Sub-Saharan Africa, as well as Western Sahara and part of Morocco. The two major mediterranean seas of the world also flow to the Atlantic:

- The Caribbean Sea and Gulf of Mexico basin includes most of the U.S. interior between the Appalachian and Rocky Mountains, a small part of the Canadian provinces of Alberta and Saskatchewan, eastern Central America, the islands of the Caribbean and the Gulf, and a small part of northern South America.
- The Mediterranean Sea basin includes much of North Africa, east-central Africa (through the Nile River), Southern, Central, and Eastern Europe, Turkey, and the coastal areas of Israel, Lebanon, and Syria.

- The Arctic Ocean drains most of Western and Northern Canada east of the Continental Divide, northern Alaska and parts of North Dakota, South Dakota, Minnesota, and Montana in the United States, the north shore of the Scandinavian peninsula in Europe, and much of central and northern Russia, and parts of Kazakhstan and Mongolia in Asia, which totals to about 17% of the world's land.
- Just over 13% of the land in the world drains to the Pacific Ocean. Its basin includes much of China, eastern and southeastern Russia, Japan, the Korean Peninsula, most of Indochina, Indonesia and Malaysia, the Philippines, all of the Pacific Islands, the northeast coast of Australia, and Canada and the United States west of the Continental Divide (including most of Alaska), as well as western Central America and South America west of the Andes.
- The Indian Ocean's drainage basin also comprises about 13% of Earth's land. It drains the eastern coast of Africa, the coasts of the Red Sea and the Persian Gulf, the Indian subcontinent, Burma, and most of Australia.
- The Southern Ocean drains Antarctica. Antarctica comprises approximately eight percent of the Earth's land.

Largest River Basins

The five largest river basins (by area), from largest to smallest, are the basins of the Amazon, the Río de la Plata, the Congo, the Nile, and the Mississippi. The three rivers that drain the most water, from most to least, are the Amazon, Ganga, and Congo rivers.

Endorheic Drainage Basins

Endorheic drainage basins are inland basins that do not drain to an ocean. Around 18% of all land drains to endorheic lakes or seas or sinks. The largest of these consists of much of the interior of Asia, which drains into the Caspian Sea, the Aral Sea, and numerous smaller lakes.

Endorheic basin in Central Asia

Other endorheic regions include the Great Basin in the United States, much of the Sahara Desert, the drainage basin of the Okavango River (Kalahari Basin), highlands near the African Great Lakes, the interiors of Australia and the Arabian Peninsula, and parts in Mexico and the Andes. Some of these, such as the Great Basin, are not single drainage basins but collections of separate, adjacent closed basins.

In endorheic bodies of standing water where evaporation is the primary means of water loss, the water is typically more saline than the oceans. An extreme example of this is the Dead Sea.

IMPORTANCE OF DRAINAGE BASINS

Geopolitical Boundaries

Drainage basins have been historically important for determining territorial boundaries, particularly in regions where trade by water has been important. For example, the English crown gave the Hudson's Bay Company a monopoly on the fur trade in the entire Hudson Bay basin, an area called Rupert's Land. Bioregional political organization today includes agreements of states (e.g., international treaties and, within the U.S.A., interstate compacts) or other political entities in a particular drainage basin to manage the body or bodies of water into which it drains. Examples of such interstate compacts are the Great Lakes Commission and the Tahoe Regional Planning Agency.

Hydrology

In hydrology, the drainage basin is a logical unit of focus for studying the movement of water within the hydrological cycle, because the majority of water that discharges from the basin outlet originated as precipitation falling on the basin. A portion of the water that enters the groundwater system beneath the drainage basin may flow towards the outlet of another drainage basin because groundwater flow directions do not always match those of their overlying drainage network. Measurement of the discharge of water from a basin may be made by a stream gauge located at the basin's outlet.

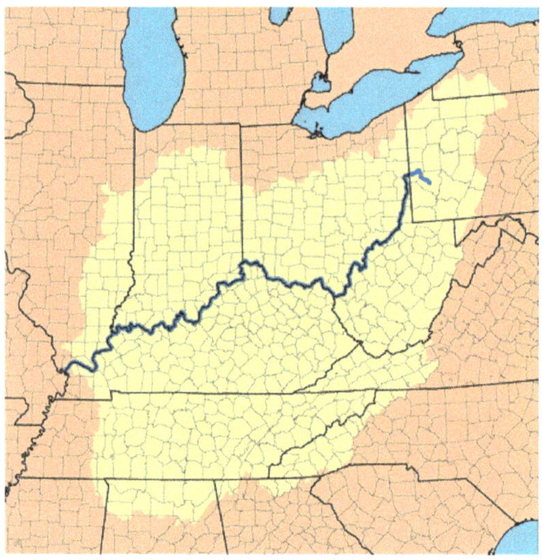

Drainage basin of the Ohio River, part of the Mississippi River drainage basin

Rain gauge data is used to measure total precipitation over a drainage basin, and there are different ways to interpret that data. If the gauges are many and evenly distributed over an area of uniform precipitation, using the arithmetic mean method will give good results. In the Thiessen polygon method, the drainage basin is divided into polygons with the rain gauge in the middle of each polygon assumed to be representative for the rainfall on the area of land included in its polygon. These polygons are made by drawing lines between gauges, then making perpendicular bisectors of those lines form the polygons. The isohyetal method involves contours of equal precipitation are drawn over the gauges on a map. Calculating the area between these curves and adding up the volume of water is time consuming.

Isochrone maps can be used to show the time taken for runoff water within a drainage basin to reach a lake, reservoir or outlet, assuming constant and uniform effective rainfall.

Geomorphology

Drainage basins are the principal hydrologic unit considered in fluvial geomorphology. A drainage basin is the source for water and sediment that moves from higher elevation through the river system to lower elevations as they reshape the channel forms.

Ecology

Drainage basins are important in ecology. As water flows over the ground and along rivers it can pick up nutrients, sediment, and pollutants. With the water, they are transported towards the outlet of the basin, and can affect the ecological processes along the way as well as in the receiving water source.

Modern use of artificial fertilizers, containing nitrogen, phosphorus, and potassium, has affected the mouths of drainage basins. The minerals are carried by the drainage basin to the mouth, and may accumulate there, disturbing the natural mineral balance. This can cause eutrophication where plant growth is accelerated by the additional material.

Resource Management

Because drainage basins are coherent entities in a hydro-logical sense, it has become common to manage water resources on the basis of individual basins. In the U.S. state of Minnesota, governmental entities that perform this function are called "watershed districts". In New Zealand, they are called catchment boards. Comparable community groups based in Ontario, Canada, are called conservation authorities. In North America, this function is referred to as "watershed management". In Brazil, the National Policy of Water Resources, regulated by Act n° 9.433 of 1997, establishes the drainage basin as the territorial division of Brazilian water management.

When a river basin crosses at least one political border, either a border within a nation or an international boundary, it is identified as a transboundary river. Management of such basins becomes the responsibility of the countries sharing it. Nile Basin Initiative, OMVS for Senegal River, Mekong River Commission are a few examples of arrangements involving management of shared river basins.

Management of shared drainage basins is also seen as a way to build lasting peaceful relationships among countries.

CATCHMENT FACTORS

The catchment is the most significant factor determining the amount or likelihood of flooding.

Catchment factors are: topography, shape, size, soil type, and land use (paved or roofed areas). Catchment topography and shape determine the time taken for rain to reach the river, while catchment size, soil type, and development determine the amount of water to reach the river.

Topography

Generally, topography plays a big part in how fast runoff will reach a river. Rain that falls in steep mountainous areas will reach the primary river in the drainage basin faster than flat or lightly sloping areas (e.g., > 1% gradient).

Shape

Shape will contribute to the speed with which the runoff reaches a river. A long thin catchment will take longer to drain than a circular catchment.

Size

Size will help determine the amount of water reaching the river, as the larger the catchment the greater the potential for flooding. It also determined on the basis of length and width of the drainage basin.

Soil type

Soil type will help determine how much water reaches the river. Certain soil types such as sandy soils are very free-draining, and rainfall on sandy soil is likely to be absorbed by the ground. However, soils containing clay can be almost impermeable and therefore rainfall on clay soils will run off and contribute to flood volumes. After prolonged rainfall even free-draining soils can become saturated, meaning that any further rainfall will reach the river rather than being absorbed by the ground. If the surface is impermeable the precipitation will create surface run-off which will lead to higher risk of flooding; if the ground is permeable, the precipitation will infiltrate the soil.

Land use

Land use can contribute to the volume of water reaching the river, in a similar way to clay soils. For example, rainfall on roofs, pavements, and roads will be collected by rivers with almost no absorption into the groundwater.

REFERENCES

1. Lambert, David (1998). The Field Guide to Geology. Checkmark Books. pp. 130–13. ISBN 0-8160-3823-6.
2. "Recommended Watershed Terminology". watershed.org. Retrieved 2011-08-26.
3. "What is a watershed and why should I care?". university of delaware. Retrieved 2008-02-11.
4. "Hydrologic Unit Geography". Virginia Department of Conservation & Recreation. Retrieved 21 November 2010.
5. "drainage basin". www.uwsp.edu. Retrieved 2008-02-21.
6. Encarta Encyclopedia articles on Amazon River, Congo River, and Ganges Published by Microsoft in computers.
7. Bell, V. A.; Moore, R. J. (1998). "A grid-based distributed flood forecasting model for use with weather radar data: Part 1. Formulation" (PDF). Hydrology and Earth System Sciences. Copernicus Publications. 2: 265–281. doi:10.5194/hess-2-265-1998.

8. Subramanya, K (2008). Engineering Hydrology. Tata McGraw-Hill. p. 298. ISBN 0-07-064855-7.
9. "EN 0705 isochrone map". UNESCO. Retrieved March 21, 2012.
10. "Isochrone map". Webster's Online Dictionary. Retrieved March 21, 2012.
11. http://www.strategicforesight.com/inner-articles.php?id=310#.WA9Ia-V97GI
12. DeBarry,Paul A. (2004). Watersheds: Processes, Assessment and Management. John Wiley & Sons.

CHAPTER 11

Hydrological Transport Model

Chapter Contents

- **Introduction**
- **History**
- **Types of Hydrological Transport Models**
- **Model Components**
- **Storm Water Management Model**

INTRODUCTION

An hydrological transport model is a mathematical model used to simulate river or stream flow and calculate water quality parameters. These models generally came into use in the 1960s and 1970s when demand for numerical forecasting of water quality was driven by environmental legislation, and at a similar time widespread access to significant computer power became available. Much of the original model development took place in the United States and United Kingdom, but today these models are refined and used worldwide.

River in Madagascar relatively free of sediment load

There are dozens of different transport models that can be generally grouped by pollutants addressed, complexity of pollutant sources, whether the model is steady state or dynamic, and time period modeled. Another important designation is whether the model is distributed (i.e. capable of predicting multiple points within a river) or lumped. In a basic model, for example, only one pollutant might be addressed from a simple point discharge into the receiving waters. In the most complex of models, various line source inputs from surface runoff might be added to multiple point sources, treating a variety of chemicals plus sediment in a dynamic environment including vertical river stratification and interactions of pollutants with in-stream biota. In addition watershed groundwater may also be included. The model is termed "physically based" if its parameters can be measured in the field.

Often models have separate modules to address individual steps in the simulation process. The most common module is a subroutine for calculation of surface runoff, allowing variation in land use type, topography, soil type, vegetative cover, precipitation and land management practice (such as the application rate of a fertilizer). The concept of hydrological modeling can be extended to other environments such as the oceans, but most commonly the subject of a river watershed is generally implied.

HISTORY

In 1850, T. J. Mulvany was probably the first investigator to use mathematical modeling in a stream hydrology context, although there was no chemistry involved. By 1892 M.E. Imbeau had conceived an event model to relate runoff to peak rainfall, again still with no chemistry. Robert E. Horton's seminal work on surface runoff along with his coupling of quantitative treatment of erosion laid the groundwork for modern chemical transport hydrology.

TYPES OF HYDROLOGICAL TRANSPORT MODELS

Physically Based Models

Physically based models (sometimes known as deterministic, comprehensive or process-based models) try to represent the physical processes observed in the real world. Typically, such models contain representations of surface runoff, subsurface flow, evapotranspiration, and channel flow, but they can be far more complicated. "Large scale simulation experiments were begun by the U.S. Army Corps of Engineers in 1953 for reservoir management on the main stem of the Missouri River". This, and other early work that dealt with the River Nile and the Columbia River are discussed, in a wider context, in a book published by the Harvard Water Resources Seminar, that contains the sentence just quoted. Another early model that integrated many submodels for basin chemical hydrology was the Stanford Watershed Model (SWM). The SWMM (Storm Water Management Model), the HSPF (Hydrological Simulation Program - FORTRAN) and other modern American derivatives are successors to this early work.

In Europe a favoured comprehensive model is the Système Hydrologique Européen (SHE), which has been succeeded by MIKE SHE and SHETRAN. MIKE SHE is a watershed-scale physically based, spatially distributed model for water flow and sediment transport. Flow and transport processes are represented by either finite difference representations of partial differential equations or by derived empirical equations. The following principal submodels are involved:

- Evapotranspiration: Penman-Monteith formalism
- Erosion: Detachment equations for raindrop and overland flow
- Overland and Channel Flow: Saint-Venant equations of continuity and momentum
- Overland Flow Sediment Transport: 2D total sediment load conservation equation
- Unsaturated Flow: Richards equation
- Saturated Flow: Darcy's law and the mass conservation of 2D laminar flow
- Channel Sediment Transport 1D mass conservation equation.

This model can analyze effects of land use and climate changes upon in-stream water quality, with consideration of groundwater interactions.

Worldwide a number of basin models have been developed, among them RORB (Australia), Xinanjiang (China), Tank model (Japan), ARNO (Italy), TOPMODEL (Europe), UBC (Canada) and HBV (Scandinavia), MOHID Land (Portugal). However, not all these models have a chemistry component. Generally speaking, SWM, SHE and TOPMODEL have the most comprehensive stream chemistry treatment and have evolved to accommodate the latest data sources including remote sensing and geographic information system data.

In the United States, the Corps of Engineers, Engineer Research and Development Center in conjunction with a researchers at a number of universities have developed the Gridded Surface/Subsurface Hydrologic Analysis GSSHA model. GSSHA is widely used in the U.S. for research and analysis by U.S. Army Corps of Engineers districts and larger consulting companies to compute flow, water levels, distributed erosion, and sediment delivery in complex engineering designs. A distributed nutrient and contaminant fate and transport component is undergoing testing. GSSHA input/output processing and interface with GIS is facilitated by the Watershed Modeling System (WMS).

Another model used in the United States and worldwide is V*flo*, a physics-based distributed hydrologic model developed by Vieux & Associates, Inc. V*flo* employs radar rainfall and GIS data to compute spatially distributed overland flow and channel flow. Evapotranspiration, inundation, infiltration, and snowmelt modeling capabilities are included. Applications include civil infrastructure operations and maintenance, stormwater prediction and emergency management, soil moisture monitoring, land use planning, water quality monitoring, and others.

Stochastic Models

These models based on data are black box systems, using mathematical and statistical concepts to link a certain input (for instance rainfall) to the model output (for instance runoff). Commonly used techniques are regression, transfer functions, neural networks and system identification. These models are known as stochastic hydrology models. Data based models have been used within hydrology to simulate the rainfall-runoff relationship, represent the impacts of antecedent moisture and perform real-time control on systems.

MODEL COMPONENTS

Surface Runoff Modelling

A key component of a hydrological transport model is the surface runoff element, which allows assessment of sediment, fertilizer, pesticide and other chemical contaminants. Building on the work of Horton, the unit hydrograph theory was developed by Dooge in 1959. It required the presence of the National Environmental Policy Act and kindred other national legislation to provide the impetus to integrate water chemistry to hydrology model protocols. In the early 1970s the U.S. Environmental Protection Agency (EPA) began sponsoring a series of water quality models in response to the Clean Water Act. An example of these efforts was developed at the Southeast Water Laboratory, one of the first attempts to calibrate a surface runoff model with field data for a variety of chemical contaminants.

The attention given to surface runoff contaminant models has not matched the emphasis on pure hydrology models, in spite of their role in the generation of stream loading contaminant data. In the United States the EPA has had difficulty interpreting diverse proprietary contaminant models and has to develop its own models more often than conventional resource agencies, who, focused on flood forecasting, have had more of a centroid of common basin models.

Columbia River, which has surface runoff from agriculture and logging

Example Applications

Liden applied the HBV model to estimate the riverine transport of three different substances, nitrogen, phosphorus and suspended sediment in four different countries: Sweden, Estonia, Bolivia and Zimbabwe. The relation between internal hydrological model variables and nutrient transport was assessed. A model for nitrogen sources was developed and analysed in comparison with a statistical method. A model for suspended sediment transport in tropical and semi-arid regions was developed and tested. It was shown that riverine total nitrogen could be well simulated in the Nordic climate and riverine suspended sediment load could be estimated fairly well in tropical and semi-arid climates. The HBV model for material transport generally estimated material transport loads well. The main conclusion of the study was that the HBV model can be used to predict material transport on the scale of the drainage basin during stationary conditions, but cannot be easily generalized to areas not specifically calibrated. In a different work, Castanedo et al. applied an evolutionary algorithm to automated watershed model calibration.

Lake Tahoe, headwater sub-basin of the Truckee River watershed

The United States EPA developed the DSSAM Model to analyze water quality impacts from land use and wastewater management decisions in the Truckee River basin, an area which include the cities of Reno and Sparks, Nevada as well as the Lake Tahoe basin. The model satisfactorily predicted nutrient, sediment and dissolved oxygen parameters in the river. It is based on a pollutant loading metric called "Total Daily Maximum Load" (TDML). The success of this model contributed to the EPA's commitment to the use of the underlying TDML protocol in EPA's national policy for management of many river systems in the United States.

The DSSAM Model is constructed to allow dynamic decay of most pollutants; for example, total nitrogen and phosphorus are allowed to be consumed by benthic algae in each time step, and the algal communities are given a separate population dynamic in each river reach (e.g. based upon river temperature). Regarding stormwater runoff in Washoe County, the specific elements within a new xeriscape ordinance were analyzed for efficacy using the model. For the varied agricultural uses in the watershed, the model was run to understand the principal sources of impact, and management practices were developed to reduce in-river pollution. Use of the model has specifically been conducted to analyze survival of two endangered species found in the Truckee River and Pyramid Lake: the Cui-ui sucker fish (endangered 1967) and the Lahontan cutthroat trout (threatened 1970).

STORM WATER MANAGEMENT MODEL

The United States Environmental Protection Agency (EPA) Storm Water Management Model (SWMM) is a dynamic rainfall–runoff–subsurface runoff simulation model used for single-event to long-term (continuous) simulation of the surface/subsurface hydrology quantity and quality from primarily urban/suburban areas. The hydrology component of SWMM operates on a collection of subcatchment areas divided into impervious and pervious areas with and without depression storage to predict runoff and pollutant loads from precipitation, evaporation and infiltration losses from each of the subcatchment. In addition low impact development (LID) and best management practice areas on the subcatchment can be modeled to reduce the impervious and pervious runoff. The routing or hydraulics section of SWMM transports this water and possible associated water quality constituents through a system of closed pipes, open channels, storage/treatment devices, ponds, storages, pumps, orifices, weirs, outlets, outfalls and other regulators. SWMM tracks the quantity and quality of the flow generated within each subcatchment, and the flow rate, flow depth, and quality of water in each pipe and channel during a simulation period composed of multiple fixed or variable time steps. The water quality constituents such as water quality constituents can be simulated from buildup on the subcatchments through washoff to a hydraulic network with optional first order decay and linked pollutant removal, best management practice and low-impact development (LID) removal and treatment can be simulated at selected storage nodes. SWMM is one of the hydrology transport models which the EPA and other agencies have applied widely throughout North America and through consultants and universities throughout the world. The latest update notes and new features can be found on the EPA website in the download section. Recently added in November 2015 was the EPA SWMM 5.1 Hydrology Manual

Program Description

The EPA storm water management model (SWMM) is a dynamic rainfall-runoff-routing simulation model used for single event or long-term (continuous) simulation of runoff quantity and quality from primarily urban areas. The runoff component of SWMM operates on a collection of subcatchment areas that receive precipitation and generate runoff and pollutant loads. The routing portion of SWMM transports this runoff through a system of pipes, channels, storage/treatment devices, pumps, and regulators. SWMM tracks the quantity and quality of runoff generated within each subcatchment, and the flow rate, flow depth, and quality of water in each pipe and channel during a simulation period divided into multiple time steps.

SWMM accounts for various hydrologic processes that produce runoff from urban areas. These include:

- time-varying rainfall
- evaporation of standing surface water
- snow accumulation and melting
- rainfall interception from depression storage
- infiltration of rainfall into unsaturated soil layers
- percolation of infiltrated water into groundwater layers
- interflow between groundwater and the drainage system
- nonlinear reservoir routing of overland flow
- capture and retention of rainfall/runoff with various types of low impact development (LID) practices.

SWMM also contains a flexible set of hydraulic modeling capabilities used to route runoff and external inflows through the drainage system network of pipes, channels, storage/treatment units and diversion structures. These include the ability to:

- handle networks of unlimited size·
- use a wide variety of standard closed and open conduit shapes as well as natural channels·
- model special elements such as storage/treatment units, flow dividers, pumps, weirs, and orifices·
- apply external flows and water quality inputs from surface runoff, groundwater interflow, rainfall-dependent infiltration/inflow, dry weather sanitary flow, and user-defined inflows
- utilize either kinematic wave or full dynamic wave flow routing methods·
- model various flow regimes, such as backwater, surcharging, reverse flow, and surface ponding·
- apply user-defined dynamic control rules to simulate the operation of pumps, orifice openings, and weir crest levels.

Spatial variability in all of these processes is achieved by dividing a study area into a collection of smaller, homogeneous subcatchment areas, each containing its own fraction of pervious and impervious sub-areas. Overland flow can be routed between sub-areas, between subcatchments, or between entry points of a drainage system.

Since its inception, SWMM has been used in thousands of sewer and stormwater studies throughout the world. Typical applications include:

- design and sizing of drainage system components for flood control
- sizing of detention facilities and their appurtenances for flood control and water quality protection·
- flood plain mapping of natural channel systems, by modeling the river hydraulics and associated flooding problems using prismatic channels·
- designing control strategies for minimizing Combined Sewer Overflow (CSO) and Sanitary Sewer Overflow (SSO)·
- evaluating the impact of inflow and infiltration on sanitary sewer overflows·
- generating non-point source pollutant loadings for waste load allocation studies·
- evaluating the effectiveness of BMPs and Subcatchment LID's for reducing wet weather pollutant loadings.Rainfall-runoff modeling of urban and rural watersheds
- hydraulic and water quality analysis of storm, sanitary, and combined sewer systems
- master planning of sewer collection systems and urban watersheds
- system evaluations associated with USEPA's regulations including NDPES permits, CMOM, and TMDL
- 1D and 2D (surface ponding) predictions of flood levels and flooding volume

EPA SWMM is public domain software that may be freely copied and distributed. The SWMM 5 public domain consists of C engine code and Delphi SWMM 5 graphical user interface code. The C code and Delphi code are easily edited and can be recompiled by students and professionals for custom features or extra output features.

History

SWMM was first developed between 1969–1971 and has undergone four major upgrades since those years. The major upgrades were: (1) Version 2 in 1973-1975, (2) Version 3 in 1979-1981, (3) Version 4 in 1985-1988 and (4) Version 5 in 2001-2004. A list of the major changes and post 2004 changes are shown in Table 1. The current SWMM edition, Version 5/5.1.011, is a complete re-write of the previous Fortran releases in the programming language C, and it can be run under Windows XP, Windows Vista, Windows 7, Windows 8, Windows 10 and also with a recomplilation under Unix. The code for SWMM5 is open source and public domain code that can be downloaded from the EPA Web Site.

EPA SWMM 5 provides an integrated graphical environment for editing watershed input data, running hydrologic, hydraulic, real time control and water quality simulations, and viewing the results in a variety of graphical formats. These include color-coded thematic drainage area maps, time series graphs and tables, profile plots, scatter plots and statistical frequency analyses.

This latest re-write of EPA SWMM was produced by the Water Supply and Water Resources Division of the U.S. Environmental Protection Agency's National Risk Management Research Laboratory with assistance from the consulting firm of CDM Inc under a Cooperative Research and Development Agreement (CRADA). SWMM 5 is used as the computational engine for many modeling packages plus components of SWMM5 are in other modeling packages. The major modeling packages that use all or some of the SWMM5 components are shown in the Vendor section. The update history of SWMM 5 from the original SWMM 5.0.001 to the current version SWMM 5.1.007 can be found at the EPA Download in the file epaswmm5_updates.txt. SWMM 5 was approved FEMA Model Approval Page in May 2005 with this note about the versions that are approved on the FEMA Approval Page SWMM 5 Version 5.0.005 (May 2005) and up for NFIP modeling. SWMM 5 is used as the computational engine for many modeling packages and some components of SWMM5 are in other modeling packages.

SWMM Conceptual Model

SWMM conceptualizes a drainage system as a series of water and material flows between several major environmental compartments. These compartments and the SWMM objects they contain include:

The Atmosphere compartment, from which precipitation falls and pollutants are deposited onto the land surface compartment. SWMM uses Rain Gage objects to represent rainfall inputs to the system. The raingage objects can use time series, external text files or NOAA rainfall data files. The Rain Gage objects can use precipitation for thousands of years. Using the SWMM-CAT Addon to SWMM5 climate change can now be simulated using modified temperature, evaporation or rainfall.

The Land Surface compartment, which is represented through one or more Subcatchment objects. It receives precipitation from the Atmospheric compartment in the form of rain or snow; it sends outflow in the form of infiltration to the Groundwater compartment and also as surface runoff and pollutant loadings to the Transport compartment. The Low Impact Development (LID) controls are part of the Subcatchments and store, infiltrate or evaporate the runoff.

The Groundwater compartment receives Infiltration (hydrology) from the Land Surface compartment and transfers a portion of this inflow to the Transport compartment. This compartment is modeled using Aquifer objects. The connection to the Transport compartment can be either a static boundary or a dynamic depth in the channels. The links in the Transport compartment now also have seepage and evaporation.

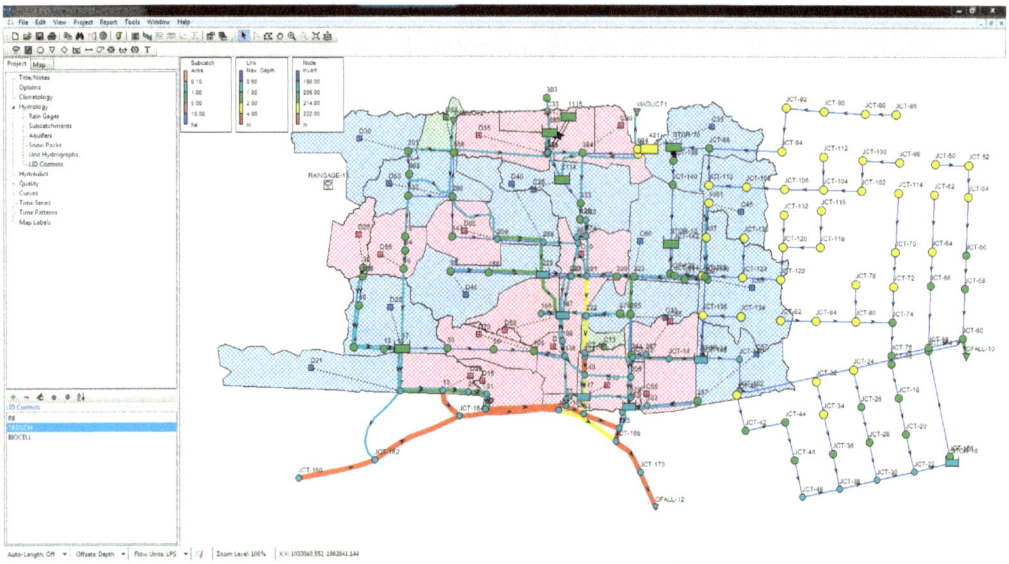

SWMM 5 Model Simulation GUI

The Transport compartment contains a network of conveyance elements (channels, pipes, pumps, and regulators) and storage/treatment units that transport water to outfalls or to treatment facilities. Inflows to this compartment can come from surface runoff, groundwater interflow, sanitary dry weather flow, or from user-defined hydrographs. The components of the Transport compartment are modeled with Node and Link objects.

Not all compartments need appear in a particular SWMM model. For example, one could model just the transport compartment, using pre-defined hydrographs as inputs. If you use the kinematic wave routing then the nodes do not need to contain an outfall.

Model Parameters

The simulated model parameters for subcatchments are surface roughness, depression storage, slope, flow path length; for Infiltration: Horton: max/min rates and decay constant; Green-Ampt: hydraulic conductivity, initial moisture deficit and suction head; Curve Number: NRCS (SCS) Curve number; All: time for saturated soil to fully drain; for Conduits: Manning's roughness; for Water Quality: buildup/washoff function coefficients, first order decay coefficients, removal equations. A study area can be divided into any number of individual subcatchments, each of which drains to a single point. Study areas can range in size from a small portion of a single lots up to thousands of acres. SWMM uses hourly or more frequent rainfall data as input and can be run for single events or in continuous fashion for any number of years.

Hydrology and Hydraulics Capabilities

SWMM 5 accounts for various hydrologic processes that produce surface and subsurface runoff from urban areas. These include:

- Time-varying rainfall for an unlimited number of raingages for both design and continuous hyetographs
- evaporation of standing surface water on watersheds and surface ponds
- snowfall accumulation, plowing and melting
- rainfall interception from depression storage in both impervious and pervious areas
- infiltration of rainfall into unsaturated soil layers
- percolation of infiltrated water into groundwater layers
- interflow between groundwater and pipes and ditches
- nonlinear reservoir routing of watershed overland flow.

Spatial variability in all of these processes is achieved by dividing a study area into a collection of smaller, homogeneous watershed or subcatchment areas, each containing its own fraction of pervious and impervious sub-areas. Overland flow can be routed between sub-areas, between subcatchments, or between entry points of a drainage system.

SWMM also contains a flexible set of hydraulic modeling capabilities used to route runoff and external inflows through the drainage system network of pipes, channels, storage/treatment units and diversion structures. These include the ability to:

- Simulate drainage networks of unlimited size
- use a wide variety of standard closed and open conduit shapes as well as natural or irregular channels
- model special elements such as storage/treatment units, outlets, flow dividers, pumps, weirs, and orifices
- apply external flows and water quality inputs from surface runoff, groundwater interflow, rainfall-dependent infiltration/inflow, dry weather sanitary flow, and user-defined inflows
- utilize either steady, kinematic wave or full dynamic wave flow routing methods
- model various flow regimes, such as backwater, surcharging, pressure, reverse flow, and surface ponding
- apply user-defined dynamic control rules to simulate the operation of pumps, orifice openings, and weir crest levels

Infiltration is the process of rainfall penetrating the ground surface into the unsaturated soil zone of pervious subcatchments areas. SWMM5 offers four choices for modeling infiltration:

Classical Infiltration Method

This method is based on empirical observations showing that infiltration decreases exponentially from an initial maximum rate to some minimum rate over the course of a long rainfall event. Input parameters required by this method include the maximum and minimum infiltration rates, a decay coefficient that describes how fast the rate decreases over time, and the time it takes a fully saturated soil to completely dry (used to compute the recovery of infiltration rate during dry periods).

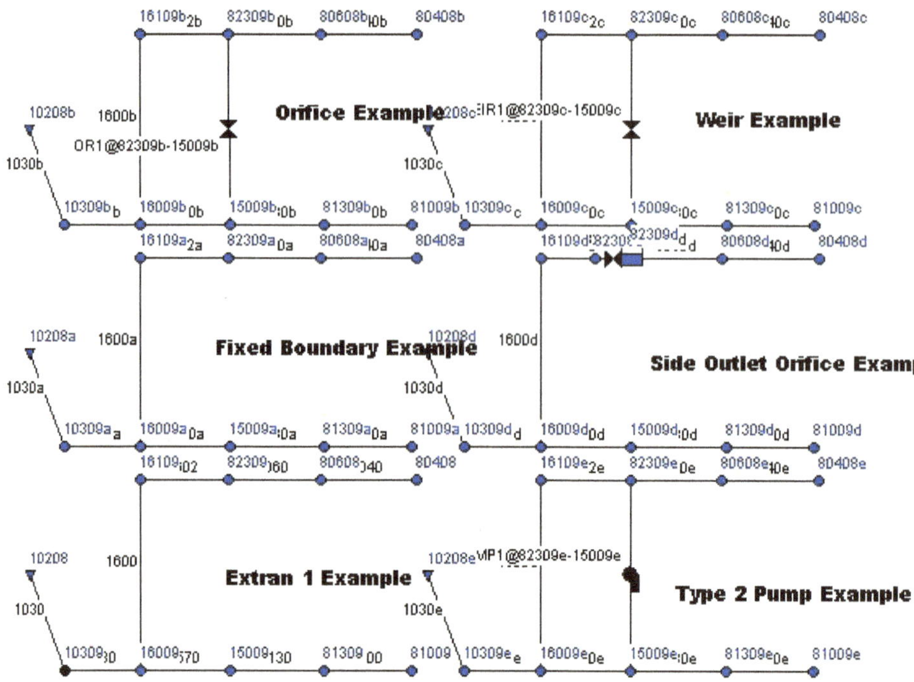

SWMM 5's QA/QC Master Example Network. This one network includes examples 1 through 7 from the SWMM 3 and SWMM 4 Manuals

Modified Horton Method

This is a modified version of the classical Horton Method that uses the cumulative infiltration in excess of the minimum rate as its state variable (instead of time along the Horton curve), providing a more accurate infiltration estimate when low rainfall intensities occur. It uses the same input parameters as does the traditional Horton Method.

Green–Ampt method

This method for modeling infiltration assumes that a sharp wetting front exists in the soil column, separating soil with some initial moisture content below from saturated soil above. The input parameters required are the initial moisture deficit of the soil, the soil's hydraulic conductivity, and the suction head at the wetting front. The recovery rate of moisture deficit during dry periods is empirically related to the hydraulic conductivity.

Curve number method

This approach is adopted from the NRCS (SCS) curve number method for estimating runoff. It assumes that the total infiltration capacity of a soil can be found from the soil's tabulated curve number. During a rain event this capacity is depleted as a function of cumulative rainfall and remaining capacity. The input parameters for this method are the curve number and the time it takes a fully saturated soil to completely dry (used to compute the recovery of infiltration capacity during dry periods).

SWMM also allows the infiltration recovery rate to be adjusted by a fixed amount on a monthly basis to account for seasonal variation in such factors as evaporation rates and groundwater levels. This optional monthly soil recovery pattern is specified as part of a project's evaporation data.

In addition to modeling the generation and transport of runoff flows, SWMM can also estimate the production of pollutant loads associated with this runoff. The following processes can be modeled for any number of user-defined water quality constituents:

- Dry-weather pollutant buildup over different land uses
- pollutant washoff from specific land uses during storm events
- direct contribution of wet and dry rainfall deposition
- reduction in dry-weather buildup due to street cleaning
- reduction in washoff load due to BMPs and LIDs
- entry of dry weather sanitary flows and user-specified external inflows at any point in the drainage system
- routing of water quality constituents through the drainage system
- reduction in constituent concentration through treatment in storage units or by natural processes in pipes and channels.

Rain Gages in SWMM5 supply precipitation data for one or more subcatchment areas in a study region. The rainfall data can be either a user-defined time series or come from an external file. Several different popular rainfall file formats currently in use are supported, as well as a standard user-defined format. The principal input properties of rain gages include:

- rainfall data type (e.g., intensity, volume, or cumulative volume)
- recording time interval (e.g., hourly, 15-minute, etc.)
- source of rainfall data (input time series or external file)
- name of rainfall data source

The other principal input parameters for the subcatchments include:

- assigned rain gage
- outlet node or subcatchment and routing fraction
- assigned land uses
- tributary surface area
- imperviousness and zero percent imperviousness
- slope
- characteristic width of overland flow
- Manning's n for overland flow on both pervious and impervious areas
- depression storage in both pervious and impervious areas
- percent of impervious area with no depression storage.

- infiltration parameters
- snowpack
- groundwater parameters
- LID parameters for each LID Control Used

Routing Options

Steady-flow routing represents the simplest type of routing possible (actually no routing) by assuming that within each computational time step flow is uniform and steady. Thus it simply translates inflow hydrographs at the upstream end of the conduit to the downstream end, with no delay or change in shape. The normal flow equation is used to relate flow rate to flow area (or depth).

This type of routing cannot account for channel storage, backwater effects, entrance/exit losses, flow reversal or pressurized flow. It can only be used with dendritic conveyance networks, where each node has only a single outflow link (unless the node is a divider in which case two outflow links are required). This form of routing is insensitive to the time step employed and is really only appropriate for preliminary analysis using long-term continuous simulations. Kinematic wave routing solves the continuity equation along with a simplified form of the momentum equation in each conduit. The latter requires that the slope of the water surface equal the slope of the conduit.

The maximum flow that can be conveyed through a conduit is the full normal flow value. Any flow in excess of this entering the inlet node is either lost from the system or can pond atop the inlet node and be re-introduced into the conduit as capacity becomes available.

Kinematic wave routing allows flow and area to vary both spatially and temporally within a conduit. This can result in attenuated and delayed outflow hydrographs as inflow is routed through the channel. However this form of routing cannot account for backwater effects, entrance/exit losses, flow reversal, or pressurized flow, and is also restricted to dendritic network layouts. It can usually maintain numerical stability with moderately large time steps, on the order of 1 to 5 minutes. If the aforementioned effects are not expected to be significant then this alternative can be an accurate and efficient routing method, especially for long-term simulations.

Dynamic wave routing solves the complete one-dimensional Saint Venant flow equations and therefore produces the most theoretically accurate results. These equations consist of the continuity and momentum equations for conduits and a volume continuity equation at nodes.

With this form of routing it is possible to represent pressurized flow when a closed conduit becomes full, such that flows can exceed the full normal flow value. Flooding occurs when the water depth at a node exceeds the maximum available depth, and the excess flow is either lost from the system or can pond atop the node and re-enter the drainage system.

Dynamic wave routing can account for channel storage, backwater, entrance/exit losses, flow reversal, and pressurized flow. Because it couples together the solution for both water levels at nodes and flow in conduits it can be applied to any general network layout, even those containing multiple downstream diversions and loops. It is the method of choice for systems subjected to significant backwater effects due to downstream flow restrictions and with flow regulation via weirs and orifices. This generality comes at a price of having to use much smaller time steps, on the order of a minute or less (SWMM can automatically reduce the user-defined maximum time step as needed to maintain numerical stability).

Integrated Hydrology/Hydraulics

One of the great advances in SWMM 5 was the integration of urban/suburban subsurface flow with the hydraulic computations of the drainage network. This advance is a tremendous improvement over the separate subsurface hydrologic and hydraulic computations of the previous versions of SWMM because it allows the modeler to conceptually model the same interactions that occur physically in the real open channel/shallow aquifer environment.

SWMM 5's LID processes include unlimited low-impact development or BMP objects per subcatchment and 5 types of layers.

The SWMM 5 numerical engine calculates the surface runoff, subsurface hydrology and assigns the current climate data at either the wet or dry hydrologic time step. The hydraulic calculations for the links, nodes, control rules and boundary conditions of the network are then computed at either a fixed or variable time step within the hydrologic time step by using interpolation routines and the simulated hydrologic starting and ending values. The versions of SWMM 5 greater than SWMM 5.1.007 allow the modeler to simulate climate changes by globally changing the rainfall, temperature and evaporation using monthly adjustments.

An example of this integration was the collection of the disparate SWMM 4 link types in the runoff, transport and extran blocks to one unified group of closed conduit and open channel link types in SWMM 5 and a collection of node types.

Low-Impact Development Components

The low-impact development (LID) function was new to SWMM 5.0.019/20/21/22 and SWMM 5.1+ It is integrated within the subcatchment and allows further refinement of the overflows, infiltration flow and evaporation in rain barrel, swales, permeable paving, green roof, rain garden, bioretention and infiltration trench. The term Low-impact development (Canada/US) is used in Canada and the United States to describe a land planning and engineering design approach to managing stormwater runoff. In recent years many states in the US have adopted LID concepts and standards to enhance their approach to reducing the harmful potential for storm water pollution in new construction projects. LID takes many forms but can generally be thought of as an effort to minimize or prevent concentrated flows of storm water leaving a site. To do this the LID practice suggests that when impervious surfaces (concrete, etc.) are used, they are periodically interrupted by pervious areas which can allow the storm water to infiltrate (soak into the earth)

You can define a variety of sub processes in each LID in SWMM5 such as: surface, pavement, soil, storage, drainmat and drain.

Each type of LID has limitations on the type of sub process allowed by SWMM 5. It has a good report feature and you can have a LID summary report in the rpt file and an external report file in which you can see the surface depth, soil moisture, storage depth, surface inflow, evaporation, surface infiltration, soil percolation, storage infiltration, surface outflow and the LID continuity error. You can have multiple LID's per subcatchment and we have had no issues having many complicated LID sub networks and processes inside the Subcatchments of SWMM 5 or any continuity issues not solvable by a smaller wet hydrology time step. The types of SWMM 5 LID compartments are: storage, underdrain, surface, pavement and soil. a bio retention cell has storage, underdrain and surface compartments. an infiltration trench lid has storage, underdrain and surface compartments. A porous pavement LID has storage, underdrain and pavement compartments. A rain barrel has only storage and underdrain compartments and a vegetative swale LID has a single surface compartment. Each type of LID shares different underlying compartment objects in SWMM 5 which are called layers.

This set of equations can be solved numerically at each runoff time step to determine how an inflow hydrograph to the LID unit is converted into some combination of runoff hydrograph, sub-surface storage, sub-surface drainage, and infiltration into the surrounding native soil. In addition to Street Planters and Green Roofs, the bio-retention model just described can be used to represent Rain Gardens by eliminating the storage layer and also Porous Pavement systems by replacing the soil layer with a pavement layer.

The surface layer of the LID receives both direct rainfall and runon from other areas. It loses water through infiltration into the soil layer below it, by evapotranspiration (ET) of any water stored in depression storage and vegetative capture, and by any surface runoff that might occur. The soil layer contains an amended soil mix that can support vegetative growth. It receives infiltration from the surface layer and loses water through ET and by percolation into the storage layer below it. The storage layer consists of coarse crushed stone or gravel. It receives percolation from the soil zone above it and loses water by either infiltration into the underlying natural soil or by outflow through a perforated pipe underdrain system.

New as of July 2013, the EPA's National Stormwater Calculator is a Windows desktop application that estimates the annual amount of rainwater and frequency of runoff from a specific site anywhere in the United States. Estimates are based on local soil conditions, land cover, and historic rainfall records. The calculator accesses several national databases that provide soil, topography, rainfall, and evaporation information for the chosen site. The user supplies information about the site's land cover and selects the types of low impact development (LID) controls they would like to use on site. The LID Control features in SWMM 5.1.007 include the following among types of Green infrastructure:

StreetPlanter: Bio-retention Cells are depressions that contain vegetation grown in an engineered soil mixture placed above a gravel drainage bed. They provide storage, infiltration and evaporation of both direct rainfall and runoff captured from surrounding areas. Street planters consist of concrete boxes filled with an engineered soil that supports vegetative growth. Beneath the soil is a gravel bed that provides additional storage. The walls of a planter extend 3 to 12 inches above the soil bed to allow for ponding within the unit. The thickness of the soil growing medium ranges from 6 to 24 inches while gravel beds are 6 to 18 inches in depth. The planter's capture ratio is the ratio of its area to the impervious area whose runoff it captures.

Main Street Tree Planter, Miles City (281991376)

Raingarden:Rain Gardens are a type of bio-retention cell consisting of just the engineered soil layer with no gravel bed below it.Rain Gardens are shallow depressions filled with an engineered soil mix that supports vegetative growth. They are usually used on individual home lots to capture roof runoff. Typical soil depths range from 6 to 18 inches. The capture ratio is the ratio of the rain garden's area to the impervious area that drains onto it.

Rain garden (2014)

GreenRoof: Green Roofs are another variation of a bio-retention cell that have a soil layer laying atop a special drainage mat material that conveys excess percolated rainfall off of the roof. Green Roofs (also known as Vegetated Roofs) are bio-retention systems placed on roof surfaces that capture and temporarily store rainwater in a soil growing medium. They consist of a layered system of roofing designed to support plant growth and retain water for plant uptake while preventing ponding on the roof surface. The thickness used for the growing medium typically ranges from 3 to 6 inches.

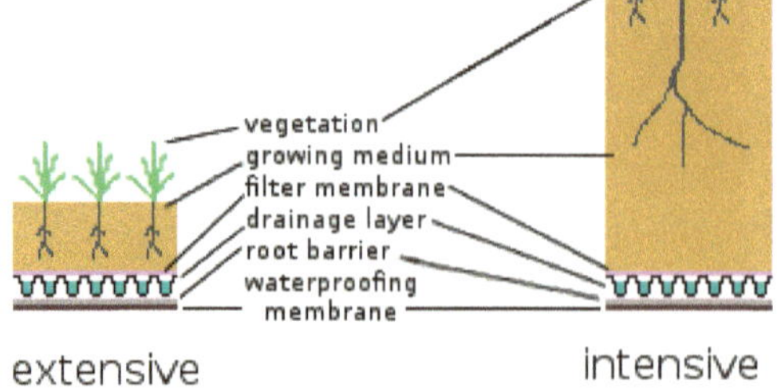

Intensive extensive green roofs

InfilTrench: infiltration trenches are narrow ditches filled with gravel that intercept runoff from upslope impervious areas. They provide storage volume and additional time for captured runoff to infiltrate the native soil below.

Infiltration trench (6438020585)

PermPave or Permeable Pavements Continuous Permeable Pavement systems are excavated areas filled with gravel and paved over with a porous concrete or asphalt mix. Continuous Permeable Pavement systems are excavated areas filled with gravel and paved over with a porous concrete or asphalt mix. Modular Block systems are similar except that permeable block pavers are used instead. Normally all rainfall will immediately pass through the pavement into the gravel storage layer below it where it can infiltrate at natural rates into the site's native soil. Pavement layers are usually 4 to 6 inches in height while the gravel storage layer is typically 6 to 18 inches high. The Capture Ratio is the percent of the treated area (street or parking lot) that is replaced with permeable pavement.

Cistern: Rain Barrels (or Cisterns) are containers that collect roof runoff during storm events and can either release or re-use the rainwater during dry periods. Rain harvesting systems collect runoff from rooftops and convey it to a cistern tank where it can be used for non-potable water uses and on-site infiltration. The harvesting system is assumed to consist of a given number of fixed-sized cisterns per 1000 square feet of rooftop area captured. The water from each cistern is withdrawn at a constant rate and is assumed to be consumed or infiltrated entirely on-site.

VegSwale: Vegetative swales are channels or depressed areas with sloping sides covered with grass and other vegetation. They slow down the conveyance of collected runoff and allow it more time to infiltrate the native soil beneath it. Infiltration basins are shallow depressions filled with grass or other natural vegetation that capture runoff from adjoining areas and allow it to infiltrate into the soil.

Wet ponds are frequently used for water quality improvement, groundwater recharge, flood protection, aesthetic improvement or any combination of these. Sometimes they act as a replacement for the natural absorption of a forest or other natural process that was lost when an area is developed. As such, these structures are designed to blend into neighborhoods and viewed as an amenity.

Dry ponds temporarily stores water after a storm, but eventually empties out at a controlled rate to a downstream water body.

Sand filters generally control runoff water quality, providing very limited flow rate control. A typical sand filter system consists of two or three chambers or basins. The first is the sedimentation chamber, which removes floatables and heavy sediments. The second is the filtration chamber, which removes additional pollutants by filtering the runoff through a sand bed. The third is the discharge chamber. Infiltration trench, is a type of best management practice (BMP) that is used to manage stormwater runoff, prevent flooding and downstream erosion, and improve water quality in an adjacent river, stream, lake or bay. It is a shallow excavated trench filled with gravel or crushed stone that is designed to infiltrate stormwater though permeable soils into the groundwater aquifer.

A Vegatated filter strip is a type of buffer strip that is an area of vegetation, generally narrow and long, that slows the rate of runoff, allowing sediments, organic matter, and other pollutants that are being conveyed by the water to be removed by settling out. Filter strips reduce erosion and the accompanying stream pollution, and can be a best management practice.

Other LID like concepts around the world include sustainable drainage system (SUDS). The idea behind SUDS is to try to replicate natural systems that use cost effective solutions with low environmental impact to drain away dirty and surface water run-off through collection, storage, and cleaning before allowing it to be released slowly back into the environment, such as into water courses.

In addition the following features can also be simulated using the features of SWMM 5 (storage ponds, seepage, orifices, Weirs, seepage and evaporation from natural channels): constructed wetlands, wet ponds, dry ponds, infiltration basin, non-surface sand filters, vegetated filterstrips, vegetated filterstrip and infiltration basin. A WetPark would be a combination of wet and dry ponds and LID features. A WetPark is also considered a constructed wetland.

SWMM5 Components

The SWMM 5.0.001 to 5.1.011 main components are: rain gages, watersheds, LID controls or BMP features such as Wet and Dry Ponds, nodes, links, pollutants, landuses, time patterns, curves, time series, controls, transects, aquifers, unit hydrographs, snowmelt and shapes (Table 3). Other related objects are the types of Nodes and the Link Shapes. The purpose of the objects is to simulate the major components of the hydrologic cycle, the hydraulic components of the drainage, sewer or stormwater network and the buildup/washoff functions that allow the simulation of water quality constituents. A watershed simulation starts with a precipitation time history. SWMM 5 has many types of open and closed pipes and channels: dummy, circular, filled circular, rectangular closed, rectangular open, trapezoidal, triangular, parabolic, power function, rectangular triangle, rectangle round, modified baskethandle, horizontal ellipse, vertical ellipse, arch, eggshaped, horseshoe, gothic, catenary, semielliptical, baskethandle, semicircular, irregular, custom and force main.

The major objects or hydrology and hydraulic components in SWMM 5 are:

- GAGE rain gage
- SUBCATCH subcatchment
- NODE conveyance system node
- LINK conveyance system link
- POLLUT pollutant
- LANDUSE land use category
- TIMEPATTERN,dry weather flow time pattern
- CURVE generic table of values
- TSERIES generic time series of values
- CONTROL conveyance system control rules
- TRANSECT irregular channel cross-section
- AQUIFER groundwater aquifer
- UNITHYD RDII unit hydrograph
- SNOWMELT snowmelt parameter set
- SHAPE custom conduit shape
- LID LID treatment units

The major overall components are called in the SWMM 5 input file and C code of the simulation engine: gage, subcatch, node, link, pollut, landuse, timepattern, curve, tseries, control, transect, aquifer, unithyd, snowmelt, shape and lid. The subsets of possible nodes are: junction, outfall, storage and divider. Storage Nodes are either tabular with a depth/area table or a functional relationship between area and depth. Possible node inflows include: external_inflow, dry_weather_inflow, wet_weather_inflow, groundwater_inflow, rdii_inflow, flow_inflow, concen_inflow, and mass_inflow. The dry weather inflows can include the possible patterns: monthly_pattern, daily_pattern, hourly_pattern and weekend_pattern.

The SWMM 5 component structure allows the user to choose which major hydrology and hydraulic components are using during the simulation:

- Rainfall/runoff with infiltration options: horton, modified horton, green ampt and curve number
- RDII
- Water Quality
- Groundwater
- Snowmelt
- Flow Routing with Routing Options: Steady State, Kinematic Wave and Dynamic Wave

SWMM 3,4 to 5 Converter

The SWMM 3 and SWMM 4 converter can convert up to two files from the earlier SWMM 3 and 4 versions at one time to SWMM 5. Typically you would convert a Runoff and Transport file to SWMM 5 or a Runoff and Extran File to SWMM 5. If you have a combination of a SWMM 4 Runoff, Transport and Extran network then you will have to convert it in pieces and copy and paste the two data sets together to make one SWMM 5 data set. The x,y coordinate file is only necessary if you do not have existing x, y coordinates on the D1 line of the SWMM 4 Extran input data[set. You can use the command File=>Define Ini File to define the location of the ini file. The ini file will save your conversion project input data files and directories.

The SWMMM3 and SWMM 3.5 files are fixed format. The SWMM 4 files are free format. The converter will detect which version of SWMM is being used. The converted files can be combined using a text editor to merge the created inp files.

SWMM-CAT Climate Change AddOn

The Storm Water Management Model Climate Adjustment Tool (SWMM-CAT) is a new addition to SWMM5 (December 2014). It is a simple to use software utility that allows future climate change projections to be incorporated into the Storm Water Management Model (SWMM). SWMM was recently updated to accept a set of monthly adjustment factors for each of these time series that could represent the impact of future changes in climatic conditions. SWMM-CAT provides a set of location-specific adjustments that derived from global climate change models run as part of the World Climate Research Programme (WCRP) Coupled Model Intercomparison Project Phase 3 (CMIP3) archive (Figure 4). SWMM-CAT is a utility that adds location-specific climate change adjustments to a Storm Water Management Model (SWMM) project file. Adjustments can be applied on a monthly basis to air temperature, evaporation rates, and precipitation, as well as to the 24-hour design storm at different recurrence intervals. The source of these adjustments are global climate change models run as part of the World Climate Research Programme (WCRP) Coupled Model Intercomparison Project Phase 3 (CMIP3) archive. Downscaled results from this archive were generated and converted into changes with respect to historical values by USEPA's CREAT project (http://water.epa.gov/infrastructure/watersecurity/climate/creat.cfm).

The following steps are used to select a set of adjustments to apply to SWMM5:

1) Enter the latitude and longitude coordinates of your location if available or its 5-digit zip code. SWMM-CAT will display a range of climate change outcomes for the CMIP3 results closest to your location.
2) Select whether to use climate change projections based on either a near term or far term projection period. The displayed climate change outcomes will be updated to reflect your choice.
3) Select a climate change outcome to save to SWMM. There are three choices that span the range of outcomes produced by the different global climate models used in the CMIP3 project. The Hot/Dry outcome represents a

model whose average temperature change was on the high end and whose average rainfall change was on the lower end of all model projections. The Warm/Wet outcome represents a model whose average temperature change was on the lower end and whose average rainfall change was on the wetter end of the spectrum. The Median outcome is for a model whose temperature and rainfall changes were closest to the median of all models.

4) Click the Save Adjustments to SWMM link to bring up a dialog form that will allow you select an existing SWMM project file to save your adjustments to. The form will also allow you to select which type of adjustments (monthly temperature, evaporation, rainfall, or 24-hour design storm) to save. Conversion of temperature and evaporation units is automatically handled depending on the unit system (US or SI) detected in the SWMM file.

Figure 4. The EPA SWMM5 Climate Change Program

EPA stormwater calculator based on SWMM5

Other external programs that aid in the generation of data for the EPA SWMM 5 model include: SUSTAIN, BASINS, SSOAP and the EPA's National Stormwater Calculator (SWC) which is a desktop application that estimates the annual amount of rainwater and frequency of runoff from a specific site anywhere in the United States (including Puerto Rico). The estimates are based on local soil conditions, land cover, and historic rainfall records (Figure 5).

REFERENCES

1. Mulvany, T.J. (1850). "On the use of self registering rain and flow gauges". Proc. Institute Civ. Eng. 4 (2): 1–8.
2. M.E. Imbeau, (1892) La Durance: Regime. Crues et inundations, Ann. Ponts Chausses Mem. Doc. Ser. 3(I) 5-18
3. Horton, R.E. (1933). "The role of infiltration on the hydrologic cycle". Trans. Am. Geophys. Union. 145: 446–460.
4. Horton, R.E. (1945). "Erosional development of streams and their drainage basins: Hydrological approach to quantitative geomorphology". Bull. Geol. Soc. Am. 56: 275–330. doi:10.1130/0016-7606(1945)56[275:edosat]2.0.co;2.
5. Report on use of electronic computers for integrating reservoir operations, vol.1 DATAmatic Corporation technical reports, prepared in cooperation with Raytheon Manufacturing Company for the Missouri River Division, Corps of Engineers, U.S. Army, January, 1957
6. M.P.Barnett, Comment on the Nile Valley Calculations, Journal of the Royal Statistical Society, Series B, vol. 19, 223, 1957
7. H.A.W. Morrice and W.N. Allan, Planning for the ultimate hydraulic development of the Nile Valley, Proceedings of the Institute of Civil Engineers, 14, 101, 1959,
8. F.S. Brown, Water Resource Development -- Columbia River Basin, in Report of Meeting of Columbia Basin Inter-Agency Committee, Portland, OR, Dec. 1958
9. D.F. Manzer and M.P. Barnett, Analysis by Simulation: Programming techniques for a High-Speed Digital Computer, in Arthur Maas et al, Design of Water Resource Systems, pp. 324--390, Harvard University Press, Cambridge, MA, 1962.
10. N.H. Crawford and R.K. Linsley. Digital simulation in hydrology: Stanford Watershed Model IV, Technical Report No.39 Stanford University, Palo Alto, Ca. (1966)
11. Abbott, P.E.O'Connell; Bathurst, J.C.; Cunge, J.A.; Rasmussen, J. (1986). "An Introduction to the European System: Systeme Hydrologique Europeen (SHE)". Journal of Hydrology. 87: 61–77. doi:10.1016/0022-1694(86)90115-0.
12. Vijay P. Singh,, Computer Models of Watershed Hydrology, Water Resource Publications, pgs. 563-594 (1995)
13. Downer, C.W., and F.L. Ogden, 2006, Gridded Surface Subsurface Hydrologic Analysis (GSSHA) User's Manual, Version 1.43 for Watershed Modeling System 6.1, System Wide Water Resources Program, Coastal and Hydraulics Laboratory, U.S. Army Corps of Engineers, Engineer Research and Development Center, ERDC/CHL SR-06-1, 207 pp.
14. Downer, C.W.; Ogden, F.L. (2004). "GSSHA: A model for simulating diverse streamflow generating processes". J. Hydrol. Engrg. 9 (3): 161–174.
15. Downer, C.W., F.L. Ogden, J. M. Niedzialek, and S. Liu, 2006, Gridded Surface/ Subsurface Hydrologic Analysis (GSSHA) Model: A Model for Simulating Diverse Streamflow Producing Processes, pp. 131–159, in Watershed Models,

V.P. Singh, and D. Frevert, eds., Taylor and Francis Group, CRC Press, 637 pp.

16. "Watershed Modeling System". Aquaveo. Retrieved 19 February 2016.
17. Vieuxinc.com
18. J.C.I. Dooge, Parameterization of hydrologic processes, JSC Study Conference on Land Surface Processes in Atmospheric General Circulation Models, 243-284 (1959)
19. C.M. Hogan, Leda Patmore, Gary Latshaw, Harry Seidman et al. Computer modeling of pesticide transport in soil for five instrumented watersheds, U.S. Environmental Protection Agency Southeast Water Laboratory, Athens, Ga. by ESL Inc., Sunnyvale, California (1973)
20. Steven Grant, I K Iskandar , Contaminant Hydrology, CRC Press (2000) ISBN 1-56670-476-6
21. Rikard Liden, Conceptual Runoff Models for Material Transport Estimations, PhD dissertation, Lund University, Lund, Sweden (2000)
22. Castanedo, F.; Patricio, M.A.; Molina, J.M. "Evolutionary Computation Technique Applied to HSPF Model Calibration of a Spanish Watershed". IDEAL. 2006: 216–223. doi:10.1007/11875581_26.
23. Development of a dynamic water quality simulation model for the Truckee River, Earth Metrics Inc., Environmental Protection Agency Technology Series, Washington D.C. (1987)
24. USEPA. 1991. Guidance for water quality-based decisions: The TMDL process, EPA 440/4-91-001. U.S. Environmental Protection Agency, Office of Water, Washington, DC.

Index

L

M

O

P

R

S

T

U

W